The Proper Care of
Malawi Cichlids

TW-124

The Proper Care of
Malawi Cichlids

Mary Sweeney

Drawings by: John R. Quinn and Lisa O'Connell.

Distributed in the UNITED STATES to the Pet Trade by T.F.H. Publications, Inc., One T.F.H. Plaza, Neptune City, NJ 07753; distributed in the UNITED STATES to the Bookstore and Library Trade by National Book Network, Inc. 4720 Boston Way, Lanham MD 20706; in CANADA to the Pet Trade by H & L Pet Supplies Inc., 27 Kingston Crescent, Kitchener, Ontario N2B 2T6; Rolf C. Hagen Ltd., 3225 Sartelon Street, Montreal 382 Quebec; in CANADA to the Book Trade by Macmillan of Canada (A Division of Canada Publishing Corporation), 164 Commander Boulevard, Agincourt, Ontario M1S 3C7; in ENGLAND by T.F.H. Publications, PO Box 15, Waterlooville PO7 6BQ; in AUSTRALIA AND THE SOUTH PACIFIC by T.F.H. (Australia), Pty. Ltd., Box 149, Brookvale 2100 N.S.W., Australia; in NEW ZEALAND by Brooklands Aquarium Ltd., 5 McGiven Drive, New Plymouth, RD1 New Zealand; in the PHILIPPINES by Bio-Research, 5 Lippay Street, San Lorenzo Village, Makati, Rizal; in SOUTH AFRICA by Multipet Pty. Ltd., P.O. Box 35347, Northway, 4065, South Africa. Published by T.F.H. Publications, Inc. Manufactured in the United States of America by T.F.H. Publications, Inc.

Contents

Introduction

Mbuna are cichlids that are found only in one lake in the world—Lake Malawi. Before discussing the mbuna, it is of value to understand where they come from and what makes their environment so unique in all the world. About 15 million years ago, in the Cenozoic era, massive earth movements, called tectonic movements, formed the rift valleys, great splits in the crust of the earth. The formation of the rift valley in Africa culminated in the filling of the deep north basin creating Lake Malawi about 2 million years ago, and that is where the story of the mbuna begins. Geologically speaking, Lake Malawi enjoys comparatively recent origin and contains a wealth of geologic and fossil evidence of its age.

Lake Malawi is effectively cut off by the

Male aquarium specimen of *Pseudotropheus* sp. zebra Metangula. Photo by Ad Konings.

Murchison Rapids of the Shire River to the south of the lake so that the dominant fish family, Cichlidae, assumed to have evolved from the riverine forms of *Tilapia* and *Haplochromis*, has been totally isolated within the lake and has evolved to a point where the lake now contains more known cichlid species than in any other lake in the world—273 described species of cichlids compared to some 40-odd non-cichlid species.

Today, Lake Malawi is the ninth largest lake in the world with an overall area of about 12,000 square miles, and is bordered by Tanzania, Mozambique, and Malawi. The lake is long, narrow, and very deep, 2,300 feet at the deepest point. Only a relatively thin surface layer of the water contains enough oxygen to sustain higher forms of life. Because of the anaerobic (absence of free oxygen) conditions at greater depths, cichlids are restricted to about the upper 300 feet of the lake. These anaerobic conditions are the result of lack of circulation in the deeper waters and the build-up of gases poisonous to fishes.

Thumbi Island West in Lake Malawi. Photo by Dr. Herbert R. Axelrod.

An underwater photo of the mbuna habitat in Lake Malawi. Photo by Dr. Herbert R. Axelrod.

Visibility in the upper waters of the lake is remarkable. The waters are normally transparent to a depth of 50 feet, giving divers and collectors a distinct

Lake Malawi is bordered by three eastern African countries, Malawi, Mozambique, and Zambia.

advantage in their search for fishes. Areas with rocky bottoms support clearer water than those areas with muddy bottoms, but algal blooms are responsible for the majority of water clouding. Algal blooms notwithstanding, this water clarity has proven very favorable for observing the habits of the mbuna in the depths of Lake Malawi. Also, given the presence of hippos and crocodiles in the lake, it is prudent for the collectors, observers, and divers to keep a wary eye out even in the clear waters.

The mbuna are concentrated in the areas of the lake where the shoreline is very rocky. The vast shores of Lake Malawi are composed of several types of habitat. Rocky

The rocky shores of Lake Malawi.
Photo by Dr. Herbert R. Axelrod.

areas are separated one from the other by open, predator-filled areas of sand or grass. These rocky, underwater "islands" provide shelter and safety for the breeding colonies of mbuna that inhabit their own rocky territories.

Each of the littoral habitats supports its

own complement of species. The estuaries and swampy zones contain species adapted to river conditions that would not be suitable for life within the lake proper. The open, sandy, or grassy zones between the rocky areas provide little or no protection for the mbuna and are generally frequented by predators, like *Rhamphochromis* spp., that have adapted to life in these open spaces either by camouflage or by making use of other offensive and/or defensive tactics. The deeper (benthic) zones, where the amount of oxygen is diminished, are the property of fishes adapted to these

conditions, such as the catfish *Bagrus meridionalis*. When these *B. meridionalis* are brought to the surface in nets cast in

Besides the rocky zones, Lake Malawi also contains grassy areas with their own species of fishes. Photo by Dr. Herbert R. Axelrod.

deep waters, the sudden change in pressure causes expansion of the swim bladder to such an extent that the stomach of the fish is everted (turned inside out) through the mouth, revealing its diet of *utaka* (plankton-feeding haplochromine cichlids).

The pelagic, or open

Lake Malawi, Africa.

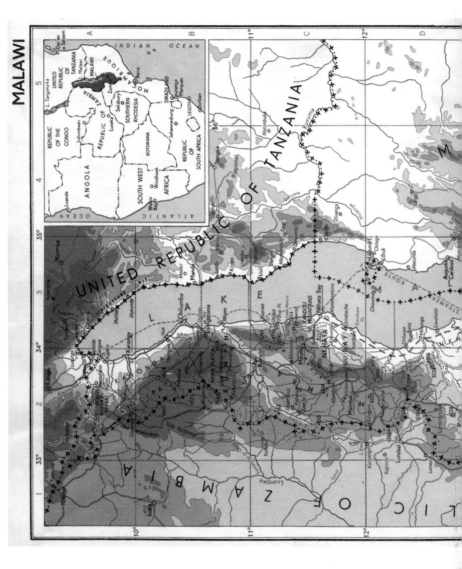

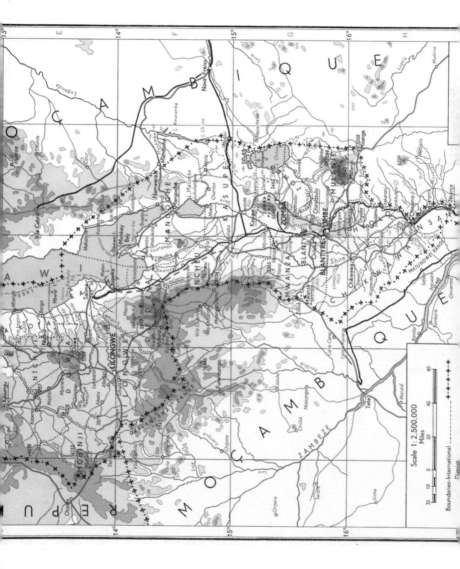

waters of Lake Malawi support the plankton feeders, which include many haplochromine cichlids and many other schooling non-cichlid fishes. The fishes of the pelagic

Haplochromis fenestratus. Photo by A. Konings.

zone are of the utmost economic importance to the local people as tens of thousands of tons of them are harvested from the lake each year to enhance the diet of the people in these areas. For clarity, it is helpful to envision the

lake as divided into these discrete zones, but in fact they often merge, as do the fish types inhabiting them. What is true, however, and of tremendous scientific importance, is that because the habitats of the rock-bound mbuna are effectively isolated by these treacherous open sandy areas, the individual populations have developed some pretty interesting idiosyncrasies. Over time, the populations have evolved in different ways. For example, if the food source of choice becomes scarce, then it becomes necessary to find an alternate. Eventually, the descendants of those

A beach seine of *Engraulicypris sardella*. Photo by A. Konings.

fishes that "learn" to utilize this alternate source of food acquire the ability to collect this new food more efficiently and perhaps this ability will become genetically fixed, thus creating a new genus or species. Even if no new genus or species evolves, the newly acquired habits will perhaps make it possible for them to colonize a new area and the isolation in this area may give rise to adaptations that may lead to the formation of a new species. What this means is that in such a vast area as Lake Malawi, with its uncounted number of isolated habitats, many such examples abound, particularly in regard to adaptations connected with methods of scraping algae and picking crustaceans from rocky surfaces. The rocky habitat of the mbuna is important not only because of the shelter from predators provided by the rocks, but also because of the algae and crustaceans that are so plentiful on the surface of the rocks. The clarity and chemical composition of the water provides favorable conditions for the growth of the life-sustaining algae. This algae and the crustaceans and other organisms that live in it have been named *aufwuchs*.

The chemical composition of the water in Lake Malawi, like all of the African rift lakes, is hard and alkaline. They differ only in degree. The pH of the water in Lake Malawi varies between 7.7 and 8.7. As a pH reading of 7.0 is neutral, the pH of the lake is considered quite alkaline. The water is also very hard. When speaking of hardness, there are two different kinds of hardness: general hardness and carbonate hardness. General hardness (total) is the measurement of the hardness caused by all the various mineral elements in the water, and is measured in

Mbuna in the lake graze constantly on algae and small crustaceans. Photo by Dr. Herbert R. Axelrod.

dGH (degrees of general hardness). Carbonate hardness is the measurement of carbonates (and bicarbonates) of calcium, magnesium, and borate in the water, and is measured in dKH (degrees of c(k)arbonate hardness). The conductivity, which is measured in microSiemens, ranges between 200 and 260. The water temperature at the shorelines where mbuna are found averages about 78°F—perfect for aquarium fishes.

Male *Pseudotropheus* sp. zebra cobalt from Nkhata Bay. Photo by A. Konings.

Pseudotropheus sp. zebra pearly. This is a natural fish, not a man-made variety! Part of what makes the mbuna so special is the palette of colors available directly from nature without specialized breeding. Photo by A. Konings.

Introducing Mbuna

WHAT ARE CICHLIDS?

The family Cichlidae is a large and remarkably diverse group of fishes that includes over 1200 species. Cichlids are representatives of the largest group of fishes, the bony fishes, Class Actinopterygii (=Osteichthys); Order Perciformes, perch-like fishes; Family Cichlidae.

Cichlids, for the most part, are perch-like, and have a laterally compressed body. Most bony fishes have two pairs of nostrils, but not the cichlids. In all cichlids, there is a single nostril on each side of the head. This feature alone is enough to distinguish them from most other fishes. There are other elements that make a cichlid a cichlid, such as incomplete scalation of the head, the single, well-developed dorsal fin, and the position and composition of the pelvic fins. The protractile mouth enables the cichlid to extend its mouth into what looks to a human very much like a yawn. This ability to extend its mouth has resulted in some very interesting and complex feeding techniques. The jaws

Pseudotropheus zebra. Photo by K. Paysan.

Pseudotropheus livingstoni. Photo by Dr. Herbert R. Axelrod.

Lamprologus brichardi. Photo by H.-J. Richter.

are armed with teeth, as are the unique pharyngeal bones, which give the cichlid a distinct advantage in the capture, manipulation, and chewing of its food. The teeth are very important, not only to the cichlids, but to the scientists that study them. The size, shape, number of surface areas, etc., all tell the scientist much about the evolution of these fishes and their method of adaptation to their environment, and make no mistake about it, cichlids are marvelous at adaptation, as you will see when we move on more specifically to mbuna.

Cichlids are well-known to aquarists for

their "intelligence" and "personality." Their intelligence manifests itself in the wild through their varied responses to the environment; in the aquarium through their recognition of the aquarist and their ability to learn tricks (and to teach the aquarist tricks).

Pseudotropheus tropheops. Photo by Dr. Herbert R. Axelrod.

Eretmodus cyanostictus. Photo by G. Meola.

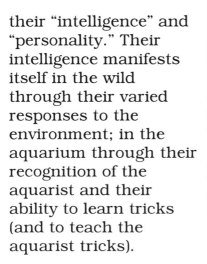

WHAT ARE MBUNA?

The term mbuna is simply the Chitonga (the language of Malawi) word for "rock fish." There are presently ten genera of mbuna: *Pseudotropheus*, *Melanochromis*, *Petrotilapia*, *Labidochromis*, *Cynotilapia*, *Labeotropheus*,

The three *Pseudotropheus* species here and on the facing page are mbunas; the *Eretmodus* and *Lamprologus* species are non-mbuna species from Lake Tanganyika.

Lamprologus lelelupi. Photo by H.-J. Richter.

Cyathochromis obliquidens. Photo by A. Konings.

Gephyrochromis, *Iodotropheus*, *Genyochromis*, and *Cyathochromis*. *Aulonocara*, although a member of the non-mbuna haplochromines, by virtue of the rock-inhabiting nature of the species often have been given the status of "honorary" mbuna.

There are many species within these genera, with more to come as scientists investigate the lake even more thoroughly. Within each species "flock" it is often impossible to distinguish specimens of closely related species because they are so similar in appearance. General anatomy, color, habitat preference, behavior, and markings are all clues to the identification of the different species, subspecies, semispecies, superspecies, sibling species, etc.

PHYSICAL CHARACTERISTICS

Mbuna are not large cichlids, which makes them very attractive as aquarium tenants. In the lake, very few mbuna reach 7.5 inches, and most of them are smaller than this. Two species barely reach a length of 3 inches. As a matter of fact, most mbuna actually grow larger in the aquarium than they do in the wild. This is unusual among aquarium fishes, where we usually find that wild fishes are much larger than their captive brethren, but the protected environment and good food in the home aquarium appears to give captive mbuna an edge in the growth department.

Mbuna have the classic cichlid shape—stocky and beautifully proportioned. Since mbuna are grazers or pickers, they are not built for rapid swimming. They are quite capable of short bursts of speed, but are usually sedate in their actions, moving back and forth and up and down with almost robot-like movements. Because they live among the rocks, they tend to be elongate and slender to facilitate their passage through tight crevices. But what really sets mbuna apart from the rest of the pack are their colors.

Camouflage is not employed as a protective device by mbuna. They are most strikingly colored and thus conspicuously draw attention to themselves. Since camouflage is widespread among fishes and works to their great advantage, there must some "reason" that mbuna are so conspicuous.

There appear to be two main functions for the gaudy coloration of

The vessel *Mbuna* docked on the shore of Lake Malawi between trips. Photo by Dr. Herbert R. Axelrod.

mbuna. One obvious reason is species recognition; coloration informs other members of the same species of their kinship. Since they so often live closely in complex communities, individual coloration is like a beacon to others of the same species.

Aulonocara sp. night. Photo by G. Medla.

The other important function of color is seen during reproduction. Apparently when color is used (as it is with mbuna) as a component of courtship, it creates the excitement necessary to complete the spawning act. If a male were to display the "wrong" color to a female of another species, the spawning cycle would be broken and

Facing page: An albino *Pseudotropheus zebra*. Photo by Dr. H. Grier.

hybridization prevented.

A final point about color is the ability of some cichlids to change it. Changes in color reflect mood, sexuality, health, aggression, and self-defense.

Mbuna are inquisitive, intelligent, and alert. These qualities are very apparent in the aquarium where they will investigate every nook and cranny in considerable detail. They are also quite aware of the goings on in their room, of their aquarist, and of potential feedings.

Pseudotropheus acii. Photo by A. Kapralski.

TERRITORIALITY AND AGGRESSION

Mbuna are for the most part territorial. They are very possessive of their space and it is extremely important to them—more important than life itself at times. It is not uncommon for two mbuna to fight to the death over a piece of rock...even when there are other perfectly suitable rocks available. But no, they will tear each other to bits before giving an inch. All this has to do with feeding and "owning" an appropriate spawning site.

Pseudotropheus zebra OB/BB morph. Dr. H. Grier.

Aulonocara hueseri. Photo by A. Konings.

Pseudotropheus tropheops spawning. Photo by A. Kapralski.

The male is very highly territorial as all spawning takes place in his territory and he must defend the spawning site from intruders. The female is very mildly territorial when in the male's territory, which is only right, since she is merely visiting.

Jaw-locking is common in mbuna, and while it looks very ferocious, it is not the worst form of fighting in these fishes. They have very tough mouths that are not easily damaged. More serious is when a fish approaches another at a right angle and bites its flank. Because cichlids are armed with formidable teeth, such attacks can inflict terrible damage. They will also chew the tails and fins off their rivals. If the victim assumes the correct submissive

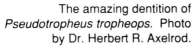

The amazing dentition of *Pseudotropheus tropheops*. Photo by Dr. Herbert R. Axelrod.

posture, the attack is inhibited, but this is often not the case.

Oddly enough, all this violence is a sign of the highly civilized status of mbuna. The value of aggression in securing a territory enhances the male's breeding success. In the aquarium, boundary fighting is common between males that occupy contiguous territories and usually consists of jaw-locking and fin-flaring.

Because of this very territorial behavior, the aquarist is often advised to utilize "controlled crowding" to avoid losses due to aggression. Controlled crowding is when you keep about twice the normal number of

Aulonocara sp. red peacock. Photo by Dr. H. Grier.

fishes in a tank. This is advisable with mbuna, but only when proper hygiene, filtration, and careful feedings are carried out.

Considering that a territory is an area occupied by a male or, rarely, a strong female, the borders of this territory depend upon the number of other fishes inhabiting the tank and the relationship of the territorial fish to the other fishes. Borders are commonly determined by natural barriers such as driftwood, rocks, large plants, and so on. If it is a large territory, then it is more stable in the sense that the same fish occupies it for a longer time. Small territories are more unstable. It can happen that the territory is a whole tank, but there are situations where several smaller territories can exist within the same aquarium. Some territories can partially overlap others, and territories can contain other territories. If the number of fishes in an aquarium is very large, say 40 mature fishes in a 50-gallon tank (which, by the way, is far too many, even by "controlled crowding" standards) with no natural borders, then it can almost be said that territories do not exist. The territory is a flexible notion, and its definition depends not only on the natural setting of the tank, but also on the number and relative strengths of its inhabitants.

"Controlled crowding" of young mbuna. Photo by H. Stolz.

The Mbuna Habitat

Since mbuna are found in clear, alkaline waters around rocks, that is the type of habitat we want to recreate for them in the aquarium. They will only thrive when offered similar conditions to those found in their native Lake Malawi. The rocky beds of Lake Malawi are composed of large, hard rock boulders and smaller, softer rocks that have been broken down by the winds and waters. The water is warm, about 78°F, and hard.

CHOOSING THE AQUARIUM

The aquarium for mbuna should be proportionate in size to the number of fishes you intend to keep. If you are interested in keeping only a breeding pair, then a 20-gallon tank will suffice. If you want a busy community tank, think in the range of 50 to 100 gallons.

Petrotilapia sp. female. Photo by Mark Smith.

The tank itself should be the standard rectangle. The fancier shapes like hexagons and cylinders just won't give you the foundation area you will need to construct your rocky territories. A tall thin tank is out of the question since mbuna are inclined to stay near the bottom and surface only reluctantly for special treats. If you

This attractive aquarium stand is very convenient with space for equipment and supplies. Photo by V. Serbin.

If you like albino fishes, these *Pseudotropheus zebra* are real eye-catchers. Photo by K. Miner.

do notice a fish hanging out in a corner of the tank near the surface, you can be pretty sure it has been harassed almost to death. It's time to move this fish to another tank.

The aquarium can be either glass or acrylic. It's strictly a matter of personal preference. Most important are size, shape, and quality. Considering the amount of rockwork employed in tanks for mbuna, quality becomes especially important. Never buy a repaired leaker, or a used tank, or a discount-store tank. Only buy a well-known brand name from a reputable pet shop. The dealer who sells African cichlids will fully understand the necessity of a quality tank and will provide the same. When purchasing an all-glass aquarium, make sure that it is not chipped, and that the edges are nicely beveled so that they will not cut you when the aquarium is handled.

The larger the tank is, the thicker the glass must be to withstand the tremendous pressure water can exert. Very big tanks will have supporting struts fitted on top from front to back. All quality tanks will

Facing page: The "Double Bubble Aquarium" is a high-tech piece of decorator aquarium furniture that will showcase your African jewels beautifully. Photo courtesy of American Acrylic Manufacturing.

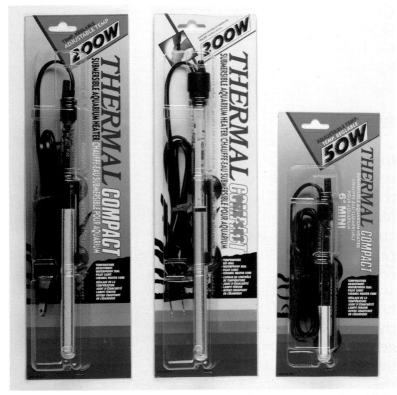

Your heater is a very important accessory in the mbuna aquarium. High-quality heaters like the Hagen Thermal Compact line will give you years of reliable service. Photo courtesy of Hagen.

feature supports for a glass cover; highly recommended for limiting the amount of dust that will fall onto the water surface, and to minimize water loss through evaporation.

Glass covers for large tanks should be hinged for access without having to remove the cover.

HEAT

Mbuna have evolved to live within a certain temperature range. The temperature in Lake Malawi is fairly constant within the range of 73 to 82°F. All fishes suffer if subjected to waters hotter or colder than that in which they have genetically evolved to live, and, regardless of their relative hardiness, all fishes will especially suffer if the temperature changes more than 3.5°F within a 12-hour period. Although a sudden rise in temperature is more

Submersible pre-set heaters can be placed parallel to the bottom of the tank. One advantage to this is that the heater won't break through accidental exposure to cool air during water changes. Photo courtesy of Hagen.

easily tolerated than a sudden drop of the same dimension, temperature changes upward create an additional hazard because as the water heats up, it is less able to hold oxygen. In a

Modern heaters are very reliable and easy to set to a stable temperature. Photo by V. Serbin.

Facing page: Head study of *Aulonocara baenschi*. Notice how the upper lip extends further than the lower lip. This facilitates grazing on rocks. Photo by M.P. & C. Piednoir.

heavily stocked aquarium this might prove to be fatal to the fishes. As the temperature rises so does the activity level of the fishes, thus when combined with a reduced oxygen level, it can be seen just how important it is to provide the preferred temperature for mbuna.

It may be possible to keep mbuna without additional heat in warm climates, but if the overnight temperature drops, it is possible that the fishes will become ill or even die, so even in warm climates it is advisable to have a heater in the tank to prevent catastrophic drops in temperature.

There are many heaters on the market, but the most popular type these days is the submersible heater controlled by a thermostat.

SPACE HEATING

Space heating is based on the idea that it is more economical to heat a room to a given temperature, and thus all aquaria in it, rather than to heat the tanks individually. It is a system that is only worthwhile if you have many tanks. Even so, this system has its flaws. The temperature must be slightly higher than required by the fishes to ensure that the water temperature is high enough. Air loses heat much faster

Male *Aulonocara stuartgranti* Mbenji. Photo by A. Spreinat.

than water does, so the heat must be kept on for long periods. In addition, tropical temperatures would be most uncomfortable for the aquarist to work in, and they might be unhealthy as well.

AQUARIUM BASE HEATING

Less costly than total space heating is base heating of the aquarium. This can be achieved in numerous ways, but the most common method is via special heating pads. These carry electric wires rather like an electric blanket, and slowly heat the tank base. The temperature can be controlled by a thermostat. The drawback is that it is still a rather wasteful way to heat water because much heat is lost to the glass, and to the gravel substrate.

Labeotropheus fuellborni. Photo by A. Roth.

IMMERSION OR SUBMERSIBLE HEATERS

The submersible heater is by far the most popular heater used in aquaria. Most contain built-in thermostats. They may be totally immersible, or may clip onto the side of the aquarium, with just the control cap above the water.

The disadvantage of having a thermostat built into a heater unit is that if it or the heater should break down you have lost both elements. Most heaters are now combined units. Some have pilot lights to indicate when they are working.

With the clip-on heaters the temperature can be

Labidochromis chisumulae. Photo by M.P. and C. Piednoir.

adjusted externally but these models are not usually as reliable or as costly as the fully submersible units. The latter, however, have the disadvantage that when adjustments are needed the heater must be unplugged and allowed to cool down for a few minutes, as it must be removed from the aquarium in order to make the adjustments.

Another type of heater now available is one that can be preset to a given temperature via a calibrated scale which is built-in. Once it is placed into the

Female *Melanochromis auratus*. Photo by K. Knaack.

tank and switched on, it will automatically heat the water to the required temperature. The clip-on heaters are not waterproof so must never be fully, submerged (only the

heater and thermostat parts).

Heaters should never be placed into the gravel; this can result in uneven heating of the glass casing, which might then shatter, or at least crack and allow water to enter the heater with obvious results. The heater can be placed into a plastic sheath that is perforated with holes. This will prevent fishes from accidentally rubbing themselves against the heater, which might burn them, depending on the power of the unit.

In large aquaria it is always better to use two heaters, one at each end of the aquarium, than one high-wattage heater.

This not only ensures more even distribution of heat, but if one heater fails then at least the temperature drop will take much longer, as it is unlikely both will fail at the same time.

Labeotropheus trewavasae.
Photo by A. Roth.

The heater should be placed so that it heats the lower levels, thus creating an upward convection current, rather than be placed where it will only heat the middle or upper waters, which will induce stratification. This will be unlikely if you are using aerators or power filters, which is probable. Of course, a heater can be placed into an external filter box, thus forming a thermofilter, but here the problem is that if

Male *Labidochromis exasperatus.*
Photo by K. Knaack.

Melanochromis auratus male.
Photo by A. Roth.

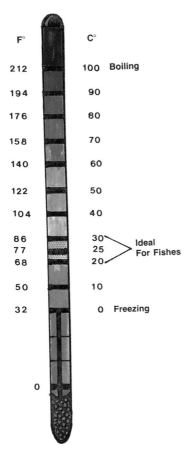

F°	C°	
212	100	Boiling
194	90	
176	80	
158	70	
140	60	
122	50	
104	40	
86	30	Ideal
77	25	For Fishes
68	20	
50	10	
32	0	Freezing
0		

Aquarium thermometers usually have a band of color indicating safe temperatures for tropical fishes.

the pump fails for any reason then the heating system is out as well.

THERMOMETERS

Even though the heater may be controlled by a thermostat, you should always use an internal or external thermometer, or both, in the aquarium. A thermostat may fail, and this will result in the water temperature rising to the full potential of the heater. This may just be enough to cook the fish! On the other hand, it might jam in the closed position so the heater does not come on. The thermometer is thus a very valuable, yet inexpensive, accessory.

There are numerous types of thermometers available, each having their own advocates. Some are free floating, some are standing, others are fixed inside the tank with suction cups, and yet others can be clipped to the sides of the aquarium. The most recent types are liquid crystal and are attached to the outside of the tank in a convenient, yet unobtrusive place. They are self-adhesive and very accurate. Do not use a mercury thermometer inside an aquarium, for should it break the poisonous contents will kill the fishes. Alcohol is a much better fluid for inside models.

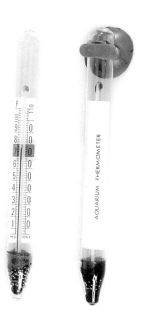

It's a big help if your floating thermometer is attached to the tank with a suction cup.

LIGHTING

Light is a most important consideration in the aquarium because it affects so many functions. The health of plants is directly linked to the amount of light they receive, for it provides the energy they need for photosynthesis. Fishes require light, for it is vital to their metabolism, and in many species it will determine their reproductive capacity. It will of course affect their color as is easily appreciated if you

Thermometers are vitally important when keeping tropical fishes. The temperature of the water should be monitored regularly. The two styles shown here are both accurate and affordable. Photo courtesy of Hagen.

consider both fishes and plants that are native to low light habitats—such as cave fishes and plants denied light artificially. In such cases they exhibit little or no color. Mbuna are found in clear, well-lit waters. Conversely, if you wanted to reduce aggression in your mbuna tank, the first thing you should do is turn off the lights.

Since mbuna are from a tropical climate, under natural conditions there will be an equal period of day and night with twelve hours in each period. This will be about 10 to 11 hours of sunlight.

Natural Light

There is no total substitute for the natural light that comes from the sun but, unfortunately, in the aquarium situation

Labidochromis mathothoi. Photo by A. Roth.

this is not without its problems. For one thing, you cannot always rely on the sun. It might not come out for days at a time. When it does, and depending on the location of an aquarium, it can result in overheating the

Labidochromis caeruleus holding fry in the buccal cavity. If you look closely, you can see a little eye peeking out of the mother's mouth. Photo by M.P. and C. Piednoir.

water. Fishes are phototropic (they tend to gravitate towards the light) and may swim at a slight angle if the source of light were always from the front or the side of the tank. One benefit of natural light is that it does encourage the growth of algae—a definite plus when keeping mbuna.

Incandescent Lights

Years ago aquarists used ordinary house bulbs of the incandescent type, and while these are still in use today, they have largely been replaced with fluorescent tubes for numerous reasons. Incandescent bulbs generate quite a lot of heat and this is wasted energy, thus costly.

Good lighting will bring out the best color in your mbuna. Fluorescent lights burn longer, cooler, and more economically than any other type of light. Photo courtesy of Energy Savers.

Also, if cooler water is splashed on a hot bulb, it is likely that the bulb will explode. They are designed for suspended operation, so that when placed in a horizontal position in an aquarium hood or canopy they are placed under stresses that they were not designed for. Few modern aquarium canopies, or hoods, are made to accommodate this type of bulb, which requires a deeper unit, so those who favor incandescent bulbs often have to improvise and make their own canopies. The advantage of these bulbs is that they are inexpensive to purchase in the short term, though more costly to operate. They are also convenient in that should one fail there are usually others readily available in the average home. When viewed in relation to their light quality, they are better than most alternatives in the red end of the spectrum but rather weak at the blue end. So, while they will help promote plant growth, there are more suitable lights now available. The heat generated by incandescent bulbs will increase the amount of evaporation of the tank's water.

Fluorescent Lights

Although more costly to install, fluorescent lamps are the most favored choice for aquarists because of their numerous advantages. They are far more efficient, producing more lumens

Pseudotropheus callainos. Photo by M.P. and C. Piednoir.

per watt than incandescents, thus lower wattage lights are needed, which in turn means lower operating costs. The average life of fluorescent tubes is at least three times that of a incandescent bulb, and can be very much more, for the life expectancy of the fluorescent tubes is based on the number of burning hours per start. It would not be unreasonable to expect about five times the life expectancy in the average aquarium when compared to an incandescent bulb. Fluorescent tubes are cold running so the risk of them shattering when splashed is very much less. Because they are slim, the hood that houses them can be slim also, which means a more attractive style is possible.

Lighting Alternatives

While fluorescent tube lights are excellent, if they have a drawback it is that their diffused light does not penetrate to the lower levels of very deep tanks, nor are they very good for creating variable lighting effects. For those wanting especially bright light, other lamps are available. Metal-halide lamps (quartz-halogen) produce very bright light but are expensive. They will need to be suspended about 12 inches above the aquarium, which

For people interested in highlighting the blue part of the color spectrum there are actinic and fluorescent combinations.

should have no canopy or hood. They are, however, very hot and expensive to operate.

Mercury vapor lamps are somewhat less expensive than the metal-halides, yet they generate considerable brightness, especially when housed in a suitable reflector. Again, they are suspended above an open top aquarium.

Sodium vapor lamps are available, but since they have a strong bias toward the red end of the spectrum their application in aquatics is limited unless they are combined with blue-biased lighting.

There are many varieties of spotlights that can be used over aquaria, and by fitting

them with various shades one can direct beams of light to specific parts of the aquarium to create stunning light-and-dark areas.

Life-span of Lights

All lights will lose a percentage of their strength as they age. This loss can exceed 20% and if considered in conjunction with even moderate cloudiness in the water, as well as increased absorption of light by encrusted glass covers, you can see how the

We've come a long way from the days when the aquarium was illuminated and heated with a kerosene lamp! These are just a few examples of the many different lighting combinations available.

amount of light reaching the plants will slowly decrease with time. Keep this in mind and replace the lights periodically. Replacement about every 9 to 12 months is suggested.

FILTRATION

The term filtration is used here in a broad sense to embrace all processes that help to maintain water quality. Filtration can thus be divided into the following types: Mechanical, chemical, biological, vegetative.

Mechanical

This is true filtration in that the water is passed through a filter medium that captures the particles of debris.

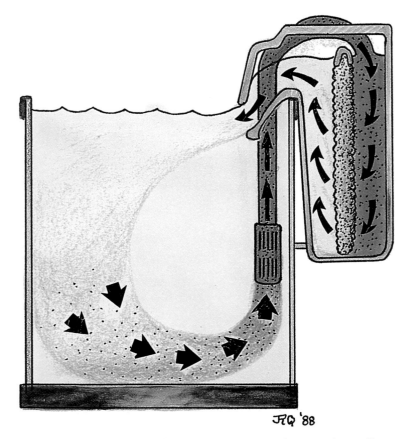

The basic principle of filtration is simply to pump the water into a filter box, canister, or whatever, so that it is forced to flow through a filter medium. The filter medium removes much of the larger suspended material from the water before the water is returned to the aquarium.

The success of such filters is dependent upon the gauge of the filter medium—the finer this is the greater its effect. Typical media are plastic strainers, nylon wool, foam cartridges, and gravel, but any inert substance can be used that will trap particles.

Chemical

Dissolved gases and chemicals will pass straight through a mechanical filter. To remove these, substances are used that will react with such chemicals and render them harmless.

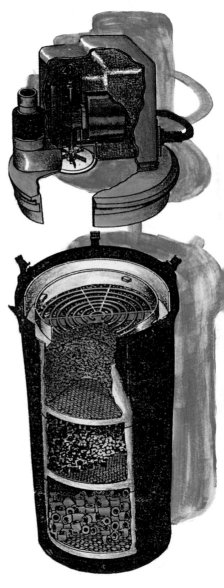

A canister filter has several separate layers of filter media and a powerful pump, designed to give optimal filtration.

The most common is activated charcoal which will adsorb many toxins such as certain heavy metals and chlorine. You must remember that a chemical filter has only done its job once the media has actually been removed from the aquarium and cleaned; until then the chemical remains a hazard. Once the media has absorbed chemicals to its saturation point then it will cease to be a chemical filter and may release the chemicals back into the water to compound the situation, so it is essential that chemical filters are cleaned on a regular basis. Most chemical filters do perform a mechanical role for they will also trap larger particles of debris, and are often used in conjunction with mechanical filters.

Biological

These filters are primarily used to encourage the nitrogen cycle to function. The media provide a home for the beneficial bacteria to colonize. The most obvious medium is the gravel substrate, but there are many other possibilities, including most mechanical filters.

The surface of the filter media is ideal for the colonization of these beneficial bacteria, and usually contain a ready source of food in the form of trapped organic matter.

Aulonocara baenschi from Benga. Photo by A. Konings.

One of the more recent mediums is sintered glass, produced in Germany, that looks somewhat like a small airstone. It is claimed to be so successful that water changes are reduced purely to that of replacing water lost in surface evaporation. Biological filters must of course have a good air supply and they do take some months to build-up the effective colony size. They are easily destroyed by

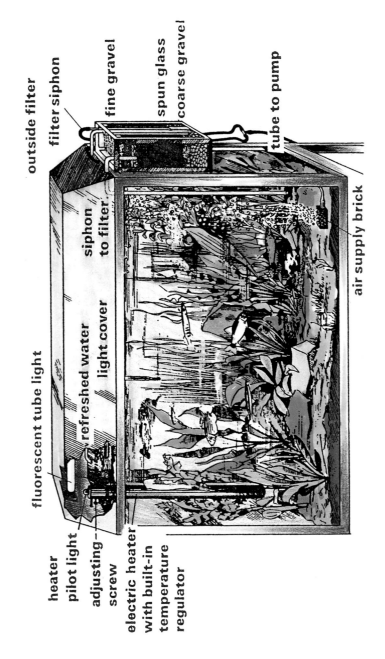

outside filter

filter siphon

fine gravel

spun glass

coarse gravel

tube to pump

air supply brick

fluorescent tube light

refreshed water

siphon to filter

light cover

heater pilot light

adjusting screw

electric heater with built-in temperature regulator

medicines added to the aquarium, so you must always be careful on this account—some medications claim not to harm such colonies.

When biological filters are cleaned this must only be done with lukewarm water, otherwise the colony will be lost and there will be a dramatic rise in the nitrite and ammonia levels. This might give the owner the impression that the problem was lack of oxygen, as the signs are much the same—the fishes gasping at the surface.

It is possible to purchase biological filter cultures from aquarium stores and these will speed up the formation of a colony.

When an aquarium is first set up there will be no colony—which is why newcomers often experience problems when they put the fishes in before the aquarium has been given a maturation period for plants and nitrifying bacteria to establish themselves.

Vegetative

Plants in an aquarium assist in cleansing and they serve as biological and chemical filters. Many fishes, and mbuna are a classic example, will quickly destroy lightly planted vegetation so you cannot have too many plants in such aquaria. However, if you wish to fully benefit from vegetative

OB morph of *Pseudotropheus zebra*.
Photo by Dr. H. Grier.

filtration then it is a matter of passing the aquarium water through another tank which is heavily planted. Another way to ensure that your plants survive with your excavation-minded mbuna is to purchase potted plants. The roots will be protected by the pot and many plant species do very well in the conditions favored by mbuna.

FILTRATION SYSTEMS

Having discussed the advantages to be gained from incorporating a filtration system, we can now look at a few practical examples and see how aeration plays an important role also.

Systems will be based on one of two principles, the simplest of which relies on an airlift to move the water through the filter media. The alternative is to incorporate a water pump in the system so that this can speed up the operation, and thus handle larger volumes. However, while it might seem advantageous to circulate the water as fast as possible, this does not always prove beneficial. Bacteria in biological filters need time to work on the organic matter, otherwise ammonia and nitrites will simply pass through the system largely untreated. Furthermore, the turbulence created in

Java fern, *Microsorium pteropus* is very good in the mbuna aquarium. It's tough and for the most part, fish leave it alone.

rapid filtration is not necessarily appreciated by mbuna and many plant species, so the turnover of water through the system must take this into consideration. As a general guide the complete aquarium water volume should pass through the filters twice in every hour—thus you simply double the tank capacity to establish the flow rate, or capacity of the pump if a power filter system is used.

Pumps are sold on the basis of how many gallons per hour they can handle. It should be noted, however, that power filter manufacturers usually quote the free flow water rate, that is, how much the pump will turn over without any obstruction. Once the filter media are placed into the system the flow rate will of course drop considerably.

Airlift Filters

The simplest filters of all are the small box filters that are made of plastic and can be situated in the corner of an aquarium. A plastic tube leads from its base and an airstone is fitted into this. A suitable filter medium, such as floss, is placed into the box, which has holes or slots through which water can flow. It is also helpful to place a layer of marbles in the bottom of the box. They will prevent the filter

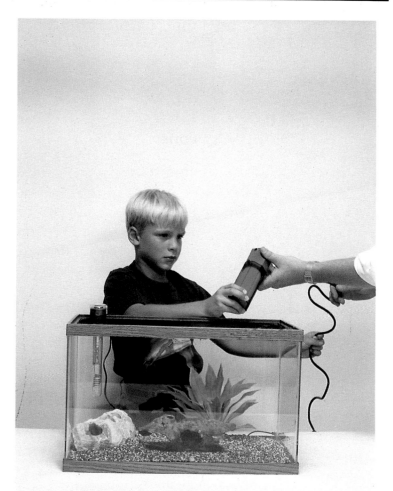

It is easiest to set up the tank first and add the water later.
Photo by I. Français.

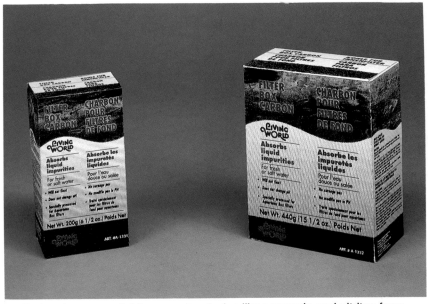

Carbon is very useful for removing toxins like ammonia and nitrites from the water. It can also be used to remove medications after treatment. Photo courtesy of Hagen.

box from floating around and provide a suitable surface area for nitrifying bacteria.

When the box is placed into the aquarium it fills with water, including the tube that opens at or just below the water surface. Once the air is switched on, it creates the upward current that draws water from the area around it, thus the debris carried into the filter collects in the box which can be regularly removed so that the filter medium can be replaced or cleaned. It is a mechanical filter and has moderate biological filtration properties.

Outside Power Filter

This is a very popular style of filter because of its efficiency, ease of maintenance, and reliability. It is simple to set up and simple to maintain. It creates both mechanical and biological filtration, and the addition of charcoal to the filter media provides the element of chemical filtration as well. Virtually every filter manufacturer offers his particular style and it remains only to select the one that suits your own taste and wallet.

Undergravel Filter

I do not recommend the undergravel filter for use with mbuna, or indeed any of the digging fishes. The undergravel filter uses a plate, usually plastic, that is raised about 1/2 inch above the bottom glass of the aquarium. The gravel covering the filter is home for the nitrifying bacteria that make this a very effective biological filter. The problem with mbuna is that they will expose the filter plate in short order due to their propensity for gravel excavation. This renders the filter virtually ineffective.

The AquaClear Aquarium Power Filter is very easy to maintain and sleek looking on the aquarium. Photo courtesy of Hagen.

This stunning yellow and ink-colored fish is called *Labidochromis caeruleus*. Photo by M.P. and C. Piednoir.

Couple this with the "dead" spots where the rocks are situated and you end up with a very inefficient, if not dangerous, undergravel filter. So please, save the U/G filter for your less construction – minded fishes.

Undergravel filters don't function well unless the gravel bed is undisturbed. The rockwork also impedes free flow of water to the filter.

Sponge Filter

The sponge filter is often overlooked, but is in reality extremely effective, particularly when raising fry. A sponge filter is coupled with a simple air pump. The water is drawn through the sponge and bubbles and water exit through a plastic stem. The surfaces of the sponge are an ideal medium for the nitrifying bacteria, collect particulate matter, and are host to rotifers that are delicious food for fishes. It is amazing to see just how quickly a cloudy aquarium is cleared when a mature sponge filter is introduced—and how clear it stays. The trick is to keep the bacteria going in the sponge. It's easy to do, just rinse the sponge periodically with lukewarm rather than hot water.

The Canister Filter

The canister filter is the workhorse of the aquarium. It seems a

Sponge filters are great if you want to save your fry. Often, the tiny fry are sucked right into other types of filters.

bit complicated in the box, but once assembled, it is really quite simple. The filter and mechanical parts are self-contained in the canister. Two hoses, one to withdraw water from the tank and one to return water to the tank, are attached to the canister. The canister is filled with layers of filter media that filter the water until it is crystal clear and biologically cleansed.

AERATION

The addition of extra air into an aquarium by mechanical means is termed aeration, and it provides many important advantages. If the air is released just above the substrate it will attract water molecules and the air/water mixture, being lighter than water, will rise to the surface. In so doing it creates a current, thus drawing water from the upper levels down. Equally important is the fact that during this same turnover, the carbon dioxide dissolved in the water is also released at the water surface. An excess of carbon dioxide is just as dangerous as a shortage of oxygen. The two factors often, but not always, go hand in hand. In an aquarium that is supersaturated with oxygen, the fishes can still suffocate if the water contains too much carbon dioxide or

You can customize your filtration by packing the Fluval Power Filter with special filter media. Photo courtesy of Hagen.

Male *Aulonocara baenschi*. Photo by A. Konings.

other poisonous gas. The fish's blood cannot rid itself of carbon dioxide quickly and therefore is unable to pick up the life-giving oxygen.

The bubbles produced when air is pumped into an aquarium contribute very little oxygen. It is the fact that they take de-oxygenated water to the surface that actually increases the overall oxygen content in the aquarium.

If the bubble size is small, and if a curtain of bubbles is produced by the airstone, then this will add some oxygen, which will be released when the bubbles burst. Not only do the air bubbles take poorly oxygenated water to the surface, but they also contribute in another way. When the bubbles burst at the surface they agitate the water, thus increasing the surface area, which

Aeration is important. Power heads, airstones, and bubble wands all contribute to the well-being of your fishes.

allows for more oxygen to dissolve into the water than would be the case if the surface was undisturbed. With maximization of oxygen content, more fishes can be housed in an aerated tank than in the same tank that is not aerated. Aeration also aids in the removal of toxic gases.

The beneficial nitrifying bacteria that convert dangerous nitrites to nitrates need a good supply of oxygen, so again aeration will have a positive effect on the aquarium's ability to remove harmful toxins.

Aeration will, of course, also reduce the risk of the water forming stratified layers, which would be the case if the water was still. It thus provides for more even distribution of heat throughout the aquarium.

Oxygen Concentration

The amount of oxygen in an aquarium will vary according to the temperature, as well as other factors. The higher the temperature is, the lower the oxygen content. You can test the water for its oxygen content by the use of a kit available from pet shops. By recording readings taken at various times, a picture can be built up of how things change over a period of time.

Pseudotropheus tropheops. Photo by A. Kapralski.

Oxygen Oversaturation

If you select only species that are known to prefer well-oxygenated waters then it would seem a good idea to pump in as much air as possible, for this would meet their needs and allow for maximum stocking levels. The risk is that if the water becomes oversaturated (supersaturated) with air it has a very negative effect. What happens is that the excess air is unable to escape into the atmosphere and so forms tiny bubbles in the water. These can often be seen on the leaves of plants. If these bubbles are taken in by the fishes as they breathe, they create problems and damage the fishes' gills. They effectively make it difficult for the fishes to breathe. Rather than help matters, the extra oxygen creates problems. The only remedy is to reduce the supply and eventually the fishes will recover. The air supply to an aquarium must therefore be balanced carefully so it provides maximum benefit, yet falls short of the point where the advantage turns to a negative.

Airstones

The normal way in which air is introduced to an aquarium is via an airstone attached by a length of plastic tubing that is

Pseudotropheus lanisticola. While not the most colorful of the mbuna, *P. lanisticola* is interesting for its habit of living in the empty shells of the snail *Lanistes*. Photo by W. Kowite.

connected to an air pump. An airstone can be any porous material, such as felt, stone, or wood; usually the molded stones are used. It may be a small cube or a long rectangular diffuser. The amount of air released can vary, not only by the pressure pumped in, but by the size of the airstone. It is usually placed near the back of the aquarium, often concealed behind a rock. By featuring two or more long airstones you can create a wall or

Airstones are inexpensive and easily attached to your pump with airline tubing.

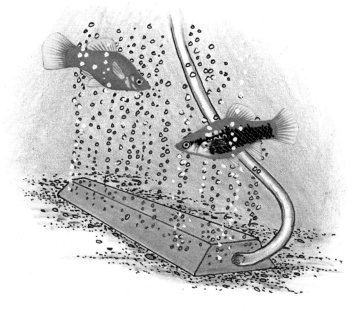

"Exotic Environments Castles" are more than decorative. The castles will be used as territories by your fish. Photo courtesy of Blue Ribbon Pet Products, Inc.

curtain of air bubbles, which can look attractive, depending on your taste.

Pumps

There is no shortage of pumps on the market and they are available in a range of outputs. By far the most popular types are those that operate by means of a diaphragm.

When the diaphragm is raised it sucks air into its chamber from an inlet chamber via an inlet valve. When the diaphragm closes it automatically closes the inlet valve and the

Air pumps power filters and airstones and it's always a good idea to have a spare on hand. Photo courtesy of Hagen.

Facing page: Which one of your clever fry will find solitude inside this urn?

air forces open the valve in the outlet chamber so that the air enters the plastic air tubing. These pumps are simple and relatively inexpensive, their only drawback being the level of noise they may produce. The better models are very quiet, but even noisier models can be made quieter by suspending them on a hook so that they are not in contact with any solid surface that would amplify the noise. It would also overcome the tendency of some of these pumps to "wander" across the unit on which they were placed. An old trick of mine is placing them on a sponge.

Diaphragm pumps are quite able to cope with two or more aquaria so that the average aquarist is unlikely to need to invest in the more robust piston pumps.

Piston pumps can from contaminating the air supply by passing the air through an air filter, and maybe also a reservoir, in order to reduce the effect of the piston's pulsating delivery of air. This is not a problem in a

deliver more air and are quieter, but they can also prove more of a problem as they need to be kept well oiled. This oil must be prevented

The diaphragm pump—a hard-working piece of aquarium equipment.

A spawning pair of *Pseudotropheus tropheops*. Photo by A. Kapralski.

diaphragm pump which operates at much greater speeds.

Placement of the Pump

An air pump must be located at least six inches above the aquarium in order that,

Dark gravel is comforting to skittish fishes. Photo by I. Français.

in the event of a sudden power outage, the water does not siphon back up the air tube and into the pump. Apart from ruining the pump, the tank could empty by such a means. To insure against this possibility, you can fit the airline with a check valve. These allow the air to pass down the tube, but close when air/water attempt to come the other way.

GRAVEL

There are probably as many kinds of gravel on the market as there are aquarium ornaments. For mbuna we want to use gravel in combination with dolomite. The dolomite contains calcium

Aquarium gravel comes in all the colors of the rainbow...and a few more. Let your taste be your guide. Photo by I. Français.

carbonate that increases the pH of the water. The best substrate is about three inches of natural-colored pebbles mixed 50/50 with the dolomite. The pebbles should be rounded and even in size. Your mbuna will delight in taking mouthfuls and moving the gravel around and piling it up

A mixed-color gravel can be very attractive in the aquarium.

For gravel-digging fishes like mbuna, stay away from substrates with sharp edges. It will damage their mouths.

Natural stone pebbles are always appropriate.

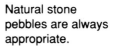

Match your drapes?

into little hills. For this reason you should never use glass particles as a substrate. Not only will it damage the fins and scales of your fish, but their mouths as well.

ROCKS

Rocks are about the most important accessory of an mbuna aquarium. Not only are they decorative, but functional as well. Mbuna really do need

This photo shows clearly the nature of this type of substrate. It is suitable only for fishes that don't dig in the gravel. Photos by I. Français.

Aquarium marbles are very decorative. Some aquarists like to create highly stylized theme aquaria and find such items useful. Photo by I. Français.

the feeling of security provided by having a cave they can call their own.

While the world is full of rocks free for the taking, do yourself and your fishes a favor and stick to safe rocks available at your pet shop. Unless you are absolutely certain (and you virtually never can be) that the rocks you find in your back yard are not going to leach

Rocks and stones are the mainstay of the classic Lake Malawi aquarium. The choice is yours. Photo by V. Serbin.

**Male *Labidochromis caeruleus*.
Photo by Dr. H. Grier.**

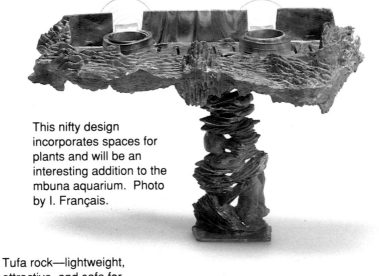

This nifty design incorporates spaces for plants and will be an interesting addition to the mbuna aquarium. Photo by I. Français.

Tufa rock—lightweight, attractive, and safe for fishes. Photo by I. Français.

Bog and drift wood are attractive and natural in the aquarium. Specimens purchased from your pet shop will also be safe and free from contaminants. Photo by I. Français.

The piece of slate attached to the bottom of this fine piece of bogwood keeps it from floating about the aquarium. Photo by I. Français.

There are many plastic decorations that are good for use with mbuna. This hollow log will be some mbuna's cherished territory. Photo by I. Français.

toxins into the water, the pet shop rocks are far cheaper when compared to the cost of replacing a tankful of fishes.

DRIFTWOOD

Driftwood is a valuable addition to the rock piles you are going to construct in your mbuna aquarium. The same caveat holds true as for rocks. Do not use driftwood you find on the beach or in your local lake. It's quite probable that it is full of unwanted passengers and/or their eggs, and you just don't know its past history.

Setting Up a Mbuna Aquarium

LOCATION

There are probably many places in your home where you could locate your aquarium. Some people even have headboard aquariums on their beds! There is no real "best" room for your new tank, but some thought should be given to the availability of electrical sockets, the water supply, and the fact that you won't want to move the aquarium after it has been set up. Kitchen, bathroom, bedroom, living room, den, or "fish room," your mbuna could care less where the tank is located as long as their needs are met. Bear in mind that you don't want to place the tank too close to windows, heat or air conditioning vents, or in an area of the home that receives particularly heavy traffic.

Another important consideration, and one that probably wouldn't occur to a first-time aquarist, is the weight of the tank. Given that each gallon of water alone weighs over eight pounds, and that 50 gallons and up are required for a community of mbuna, some thought should be given to the structural integrity of the floor. And this does not even include the weight of the tank, rocks, and interior equipment. If in doubt, check.

SET-UP OF THE TANK

You have purchased your tank and accessories and you are raring to go. You want to set up the most beautiful aquarium for your chosen mbuna. First check and make sure you have a convenient water supply. The necessary water changes can become an onerous burden and are easily neglected if you have to haul water up and down stairs or from the far end of the house. The last thing you want to do is to scrimp on those water changes! Having selected an appropriate aquarium stand, place it in your chosen location. Do not place the stand too close to the wall. There will be many occasions when you will need to work at the back of the tank and you aren't going to be able to move it easily once it's filled with water. About 8 to 12 inches between the tank and the wall should be sufficient for maintenance.

It's best to use a soft pad, like styrofoam or felt between the tank and the stand to compensate for any slight unevenness between the two that could possibly lead to a stress fracture in the glass or pop a seam.

Background coverings are very important for the mbuna's peace of mind.

Even a piece of cardboard, cut to size, is suitable.

Rinse your tank with salt and fresh water. Fill it and check for leaks. If you do find a leak at this point, it is easily repaired with silicone. If you find one after the tank is set up, you will probably have to break the whole thing down again to find and reseal the leaking seam. Set the tank on the stand.

This is the time to make sure you like the arrangement. Is the stand even? Does the tank look "right" to your eye? Any changes you have to make are much easier at this point than after you have set the whole thing up. If I sound like I'm repeating myself, it's because at this point you can save yourself a lot of headaches if you savor the work at hand rather than rushing through to get to the finished product.

Mbuna are accustomed to receiving their light from above, and in the confines of the aquarium are very nervous about light that comes from other directions, especially from below, so don't be tempted to skip the gravel substrate in favor of cleanliness. (Many fishkeepers choose bare-bottomed tanks so that they can immediately siphon off leftover food and fish wastes.) Some aquarists, in addition

Quality plastic plants like Second Nature's Plantastics are both attractive and useful in the mbuna aquarium. The fish will peck at the green algae that will soon cover the leaves. Photo courtesy of Tetra.

Start out with a diagram of your planned tank arrangement. It's much easier to make necessary modifications on paper.

Collect all your materials.

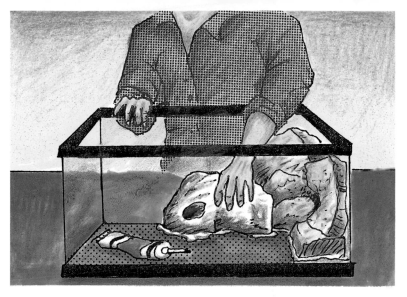

Silicone-cement the largest rocks to the bottom of the tank to give yourself a good foundation for the next levels.

The silicone must set for 24 hours before adding water to the aquarium.

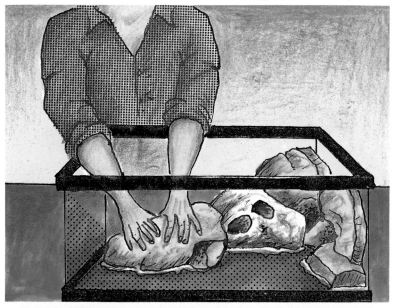

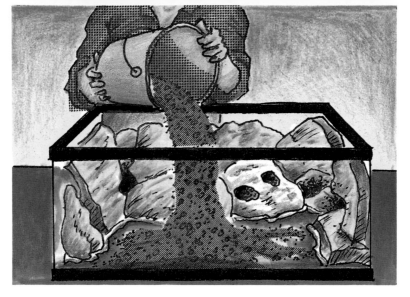

If you like the arrangement, add your well-rinsed substrate.

When you add the water, it will become very cloudy from dust left in the gravel. It's not a problem.

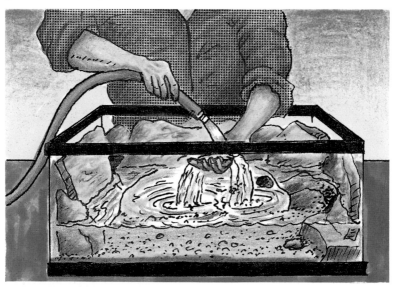

Place the remainder of your decorations and equipment.

The finished product.

to fully covering the bottom of the tank with rocks and gravel, even paint three sides with an attractive, dark paint, leaving only the front glass bare to view the fishes. The mbuna feel very secure in this kind of cave-like aquarium.

Rocks are the decorative items of choice for the mbuna tank. If you think you have enough, you will probably need more. Mbuna are fanatical about territory and love to weave in and out of any tiny nook. Given this predilection to squeeze themselves into the tightest possible places, you will want at all costs to assure that there is no danger that your rocky arrangement will ever collapse. One way to minimize the danger of collapse is to select several of the largest, smoothest rocks available and cement them to the bottom of the tank with silicone. Let the silicone dry for 24 hours and you will have a good, solid base for further construction endeavors. It helps to draw a diagram of your planned arrangement prior to using the silicone because if you have to start over it will only delay operations even further.

You won't want to set the rocks directly on the glass, because if the glass is going to crack, it will do it where a rock is creating a pressure point.

Your pet shop has very attractive rock sculptures that can only enhance your enjoyment of your beautiful aquarium. Photo by V. Serbin.

Rinse, rinse, rinse your gravel. Even after you have rinsed it enough, rinse it again. You will be amazed at how much sediment is carried in new gravel. When you fill your tank, you will see clouds of milky dust rising from the gravel. The only way to handle this is rinsing first and filtration later. You can rinse the gravel in a colander or a bucket, whichever method you prefer. Then gently pour the gravel into the tank. It is quite possible to crack your glass if you dump 30 pounds of wet gravel in all at once.

After you have your rock base and your gravel in the tank, you are then free to arrange the other rocks and driftwood to your liking, always bearing in mind that your fishes adore places with an entry and an exit. When you have to net a fish later, you will appreciate your forethought in providing exits to some of the caves. Put a net over the exit and drive the fish into the entry. Scoop up the net with the surprised fish. Sometimes it even works.

In lieu of rockwork, some aquarists are working with lengths of PVC pipe. While not quite as aesthetically pleasing, there is a lot to be said for this method. You don't have to worry about the weight of the pipe, and it is quite simple to

Top: Rocks and plants work well together. The small rocks can be used to help anchor the plants. **Bottom:** When you plan your rockwork and planting arrangements be sure to leave some open space in the center front of the aquarium for visual balance.

Top: Driftwood, rocks, and plants can be very artistically arranged with the smaller plants and rocks toward the front and the larger pieces toward the rear. **Bottom:** A top view of a workable arrangement of rock and driftwood.

modify the decor. Just move it. If you chose you could create a "condo" effect with rows of PVC sections siliconed together. The fishes love it! In essence, if it's safe for the aquarium and makes a hollow space, your mbuna are going to use it. Also, when you are designing your tank, be sure to include some spaces that are too small for your adult fishes. The fry will need plenty of havens for their first weeks until they are large enough to safely swim out in the open.

Clay flowerpots with the bottoms knocked out are ideal for the mbuna aquarium. The material is inert, or non-toxic, the color is pleasing, and, best of all, each fish can have its own pot. If you have as many pots as you do fishes, you should have very little of the nasty infighting associated with territory.

Potted plants are best around digging fishes, and mbuna are no exceptions. It's much easier to put the plants in the tank just before you add the water than to try to get just the right effect when you're up to your elbows in water. While it may at times be a bit of a challenge to keep plants with cichlids, mbuna are not as deliberately destructive when it comes to plants as many of the other cichlids (and other fishes as well), and they

This lush planting is possible if you use potted plants that the digging mbuna will not disturb. Photo by D. Van Raam.

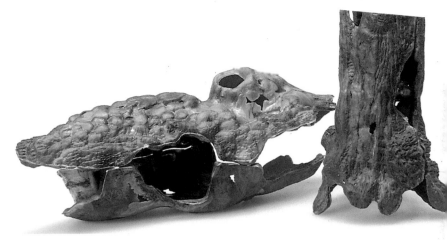

The plastic decorations sold in your pet shop are of course safe for use in the aquarium. Photo by I. Français.

do benefit from the algae often found on the leaves of living plants. For that matter, algae grow on plastic plants as well, so if you choose there is no harm in including plastic plants in your decor. Some of the many plants that do well in the alkaline conditions of the mbuna tank are: *Vallisneria* spp., *Ceratophyllum* spp., and *Potamogeton* spp.

After the tank is all set up with gravel, caves, plants, filter, and heater (but **not** turned on), and any

Ceratophyllum demersum.
Photo by R. Zukal.

Potamogeton perfoliatus.

Vallisneria portugalensis.

other equipment you have chosen, add your water. Since you have already decorated your tank, use a gentle flow of water so you don't disturb your arrangement. Do use a chlorine/chloramine remover. Most municipal water supplies must treat the water to make it safe to drink, even if these chemicals aren't good for our fishes.

Once you have added your water, it is safe to turn on your heater and filter without worrying about cracking the glass on the heater or burning out parts of the filter. Remember that it will take about 24 hours for the water temperature to stabilize, so do wait before buying your fishes.

WATER CHEMISTRY

Alkalinity is the key word when keeping mbuna. It is expressed in pH value, which literally means "hydrogen power." The pH scale ranges from 0 (extremely acidic) to 14 (extremely alkaline) with 7 being neutral. The pH value recommended for Lake Malawi cichlids is approximately 8.0.

Since the pH scale is logarithmic, the degrees on the pH scale mark a tenfold increase or decrease from the next higher or lower number, so a pH of 7 is vastly less alkaline than a pH of 8. This is not the area for approximation. An apparently small sudden change in pH is very stressful to most mbuna. While most of the mbuna available in the pet shops are tank-raised, and therefore acclimatized to somewhat less alkaline pH levels than those found in the lake, it is in their best interests to slowly raise the pH to their optimal levels once you have established them in their new homes. Ask your fish dealer the pH under which they have been kept when you purchase your fish and bear in mind how stressful sudden changes in pH are to the fishes when you introduce them to your tank.

When you have filled the aquarium with water, check the pH of your tap water. It is possible to check the pH of the water in a number of ways, depending on how often you want to check the pH and how accurate you want to be. There áre electronic pH meters on the market today where you just dip the electrode in the water and get an instant reading. These would be useful if you have a large number of tanks to test and/or are very concerned about maintaining a stable pH level. pH test

Facing page: *Pseudotropheus tropheops.*

kits are also available that are simple and quite easy to use. Either way, when keeping mbuna it is and will be necessary to know the pH of your water.

Acidic (low pH) water conditions can result from an excess of waste products in the water and are more likely to occur in an established aquarium. Brand new water is more likely to be closer to neutral, at least until the fishes are introduced. But it is always possible that the water supply in your area provides soft, acid water. What you are looking for for your mbuna is alkaline water. Fortunately this is not difficult to arrange.

The use of dolomite with your gravel as a substrate will automatically start to increase your pH as the dolomite leaches minerals into the water. Additionally, you can add sodium carbonate or trisodium phosphate to increase the alkalinity but not the hardness. Your pet shop also has special "Malawi salt" to add to the water in the amounts indicated on the label to achieve your desired pH. All in all, the proper pH level, while very important for your fishes, is not difficult to achieve.

CYCLING THE TANK

"Cycling" is a concept that is very familiar to marine aquarists and has nothing to do with bicycles. Basically, cycling a tank has to do with promoting the growth of beneficial bacteria (*Nitrobacter* and *Nitrosomonas*) in the aquarium. Without these bacteria the fish wastes and leftover food can turn a new aquarium into a toxic waste site in an amazingly short period of time. This condition in a newly set-up aquarium, manifested by almost milky, cloudy water, is known as "new tank syndrome." In an established aquarium, it is simply known as poor care.

The foundation of water quality is a process known as the "nitrogen cycle." Let's get back to the

Water conditioners like A.C.T. from Mardel Laboratories will help cleanse the water of toxic ammonia and nitrites. Photo courtesy of Mardel Laboratories.

beneficial bacteria that converts ammonia into less harmful substances in the aquarium. Ammonia is the most toxic and the most common by-product of protein metabolism, i.e. fish waste. When combined with water of a high pH, as in the Malawi aquarium, the ammonia becomes about 10 times more toxic than it would be in the same amount in an aquarium with a neutral pH. Therefore, it is critical that ammonia be kept at tolerable levels for your mbuna. While mbuna are extremely hardy fishes, ammonia poisoning is their Achilles heel, and they are very susceptible to its ill effects.

The nitrogen cycle describes the natural course of nitrogen metabolism in nature. Nitrogen is converted into ammonia as proteins break down and are acted upon by those beneficial *Nitrosomonas*. The bacterial action converts the toxic ammonia into nitrites, which in turn are acted upon by other bacteria, *Nitrobacter*, and converted into less toxic nitrates, which are then easily neutralized by plant (including algae) metabolism, zeolite and carbon in your filter, and your water changes.

Kits are available for monitoring the levels of

If you have trouble bringing soft, acidic water up to pH for Africans, try Proper pH 7.5 to help buffer the water. Photo courtesy of Aquarium Pharmaceuticals.

Male *Pseudotropheus zebra.*
Photo by A. Konings.

It is easy to maintain optimal water quality in this type of spartan, rock-filled aquarium. The bottom is painted black so as not to spook the fishes. Photo by Dr. Herbert R. Axelrod.

ammonia, nitrite, and nitrate in the aquarium and are just as important as your pH test kit if you want to keep your mbuna in top condition. You can, of course, use a hit and miss approach with water changes, but it is far more efficient, and better for your fishes, to know your ammonia level and then deal with it accordingly. Also, it is very gratifying to have your test kit tell you when all is well.

In a newly established tank, the goal is to get a good head of steam going on the culture of these beneficial bacteria, so that when your tank is at fish load capacity, you will not have the dreaded ammonia problem. One way to accomplish this, when the tank is filled and the filter running, is to introduce bacterial cultures available at the pet shop. Then you can introduce one or two inexpensive, hardy fishes like small feeder guppies. Their wastes will feed the bacteria, which in turn will cause more bacteria to establish themselves in your filter. The filter is

the home of these wonderful bacteria, and for this reason you should never clean your filter with hot water, which will kill them off.

It is almost inevitable that when you first introduce fishes to a newly set-up aquarium, your water will become cloudy. This cloudiness is normal and a manifestation of the new tank syndrome. The cloudiness shows that there is an ammonia surge, and in a few days, as the bacteria get going, the water will take on the crystal clarity all fishkeepers strive for. The absolute worst thing you can do at this point is change the water. That means you will just have to start the cycle all over again and as long as you keep changing water, the cloudiness will return.

INTRODUCING THE FISHES

This is the moment we have all been waiting for! All your hard work is about to be rewarded. You have a good filtration system, the pH is steady at 8.0, the ammonia, nitrite, and nitrate levels are 0, the temperature is 82°F. Now it is time to introduce the fishes.

When you buy your mbuna, it is likely that they will be young fish. Most pet shops carry juvenile fishes. And this is good. It is better,

especially when dealing with aggressive, territorial fishes like mbuna, to give them a chance to get adjusted to each other when they are still young and (relatively) docile. Take into account when you are stocking your tank that these little two-inch fishes will, in a short time, mature and double in size. Therefore, count on the tank looking a little bare in the beginning—but not to worry, there are fish there!

This novel "condo" style arrangement is much-loved by mbuna. They don't care that their territories aren't real rock, it's the real estate, not the geology that interests them. The fish is a tangerine morph of *Pseudotropheus zebra*. Photo by Dr. Herbert R. Axelrod.

The first thing your mbuna will do when you introduce them to your aquarium is disappear! They will immediately take full advantage of your rockwork and stake their claims to the choice hidey-holes. Don't worry, it won't be long before they associate you with dinner and reward you with an immediate audience whenever you get anywhere near the tank. Proof of their

This gorgeous tankful of various color morphs of *Pseudotropheus zebra* shows the fish off to their best advantage with light substrate, dim background, and the center spotlight. Photo by H. Stolz.

affection is that they disappear whenever a stranger approaches the tank. It could be quite embarrassing to invite a friend to see your beautiful fishes and find there isn't one in sight!

There are several ways of actually putting the fishes into the tank, but the one that I have found to work best is to put the fishes and the water they came in into a clean container. Then gradually add an equal amount of water from the aquarium. Take your time with this procedure, giving the fishes time to acclimate to the new water, then net the fishes out and put them into the aquarium. **Do not** put the water into your aquarium. Do not feed the fishes for the first day. They will probably be too interested in their new surroundings to eat and uneaten food combined with new fish waste can quickly cause water quality problems.

Leave the lights off for 24 hours to give everyone a chance to settle in. This will reduce petty squabbling and relax the fishes. You will see them soon.

A good-sized net will eliminate a lot of frustration when you have to catch your fish. Photo by I. Français.

Above: Male *Melanochromis vermivorus*. Photo by A. Roth. **Below:** Female *Melanochromis vermivorus*. Photo by A. Roth.

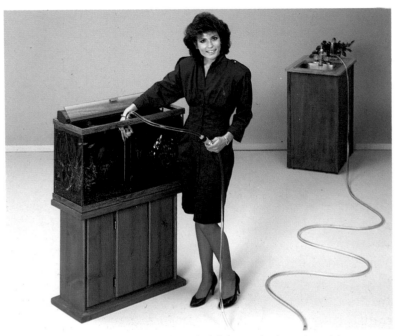

Clean water is a must for healthy, vital fishes. An investment in convenient water-changing equipment is worth every cent.

Aquarium Maintenance

When your aquarium is properly set up maintenance is a relatively simple matter. Basically, all you need to do is change 25 to 50 percent of the water every two weeks, vacuum or stir the gravel, and service the filter(s). But of course, that is a bare bones maintenance and will not do much to enhance the beauty of your aquarium.

WATER QUALITY

The most important aspect of water quality is cleanliness. All the wastes produced by your fishes, plus any decaying plant material, plus uneaten food, dead fishes, etc., all add to the pollution level of your tank. Fortunately, mbuna are very hardy and can withstand quite a bit of neglect (except for those high ammonia levels) before they show signs of stress. But we assume you want your mbuna to do more than just survive, and in this case it is important to keep the dissolved wastes in the tank to a minimum.

There are several things that contribute to the cleanliness of your aquarium water. These include keeping a reasonable population of fishes relative to the size of your tank, not overfeeding, having an adequate filtration system, using carbon and zeolite in addition to your filter media, and doing frequent partial water changes.

If you have been lax about water changes, be very careful when you do start changing the water again. Your fishes have become accustomed to the high pollution level in the tank and if you try to mend your ways by doing a massive water change all at once, you could very well kill your fishes with kindness. If you can't remember when you did your last water change, take it easy. Start by doing small water changes daily for a couple of weeks. If the fishes are tolerating this well, start to increase the amount of water changed, and decrease the frequency of changes until you feel the fishes are looking their best and the water parameters are where you want them to be.

How often and how much water you change depends on many variables: how many fishes there are, the type and amount of food offered, the temperature, etc. Just remember that small, frequent water changes are less stressful to your fishes than a large water change all at once. A 10% water change once a week is not too much, and is

Aulonocara jacobfreibergi. Photo by A. Konings.

probably a small enough amount of water not to be too much work. If you have a heavy fish load in your tank, you may need to change more water than that. For mbuna, at the very minimum, you should be changing 25% of the water every two weeks.

Water changes have an invigorating effect on your fishes. They will often trigger a reluctant, but almost-ready pair to spawn and the nice clean water will help ensure an optimal hatch rate.

When you do your water changes, be sure to match the water temperature to within a few degrees, and add a product that neutralizes chlorine and/or chloramine. Check your pH and add an appropriate amount of Malawi salts to keep the pH level stable. Remember, no sudden changes in pH.

Bear in mind that evaporation of the water in your tank will affect pH and ammonia levels. As the water evaporates, the minerals, salts, and ammonia, nitrates, and nitrites will remain in the water in greater concentrations, so it is important if you are not doing a water change to top off the tank regularly to dilute these elements.

GAS EMBOLISM

In cold climates, there is a serious problem that can occur

Pseudotropheus acii. Photo by A. Kapralski.

if you replace tank water with water straight from the tap. It is called gas embolism. You can lose every fish in your tank to this condition. Cold water has an enhanced capacity to hold dissolved gases such as chlorine, oxygen, and nitrogen. In the summer, the tepid tap water is relatively benign except for chlorine and chloramine. In the winter, this water becomes lethal with gas. When the gas-loaded water warms up in your tank (or from the addition of some warm water to bring it to tank temperature), the gas appears in the form of bubbles that cover every surface in the tank—including the fishes. The fish's gills are quite permeable to the dissolved gases and bubbles will form in the capillaries of the fishes, leading to intense pain and horrible hemorrhages. The prophylaxis is very simple—age the water for 24 hours and the gas will escape harmlessly.

DIATOM FILTER

A diatom filter is an invaluable piece of equipment for people who have problems with clouding of the water, whether it be from an algal bloom,

Facing page: *Pseudotropheus zebra* male, commonly called blueberry. Photo by Dr. H. Grier.

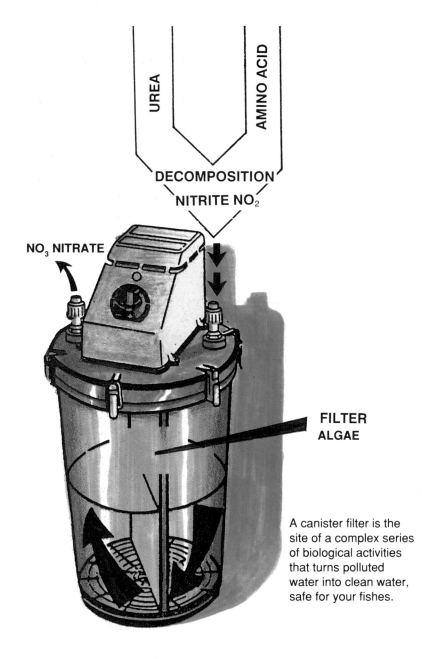

UREA

AMINO ACID

DECOMPOSITION

NITRITE NO$_2$

NO$_3$ NITRATE

FILTER ALGAE

A canister filter is the site of a complex series of biological activities that turns polluted water into clean water, safe for your fishes.

bacterial bloom, overfeeding, or whatever. In essence, the diatom filter utilizes diatomaceous earth as a filter media, straining out microscopic particles from the water and effectively "polishing" your water. A couple of hours of diatom filtering will render your aquarium water crystal clear!

BLOOMING ALGAE

Blue-green algae is a common problem in aquaria. When keeping mbuna, you want to encourage the growth of green algae, but the blue-green kind won't impress anyone, not even the mbuna. Blue-green algae is almost always a sign of anaerobic conditions in the aquarium. It often appears bright green and "greasy." It gives off a "fishy" odor. Anaerobic conditions are the result of decreased oxygenation and are commonly found in areas of the aquarium that has little or no flow of water.

Organic material gets trapped in the crevices between the grains of gravel, where it will decay, preventing the water from circulating through the gravel. The lack of circulation results in areas of the substrate where oxygen levels are low or absent. This condition will eventually become toxic to the fishes. It is therefore recommended that "gravel vacuuming" becomes part of the routine maintenance of every aquarium.

GRAVEL VACUUM

Changing water alone is not enough to keep your aquarium clean. You will be amazed at how much dirt (detritus) can collect in the substrate. If you were to take a stick and stir the gravel in an established aquarium, you probably wouldn't be able to see the fishes! Every bit of leftover food and fish waste will find its way into the crevices between the stones to decompose at leisure. There is a certain amount of bacterial activity going on in the substrate to help break down this waste, but it's never enough. You will have to do your best to maintain a healthy substrate. This includes vacuuming the gravel with a device called a "gravel washer," a long, thick tube that literally sucks the waste out of the gravel. These devices are great for water changes. There are even devices that hook right up to your faucet, clean the gravel, remove the old water, drain the waste water down the sink, and add the new water right from the tap.

FILTER MEDIA

Your choice of filter will, in large part, determine which type of filter media you will use. Whether floss and

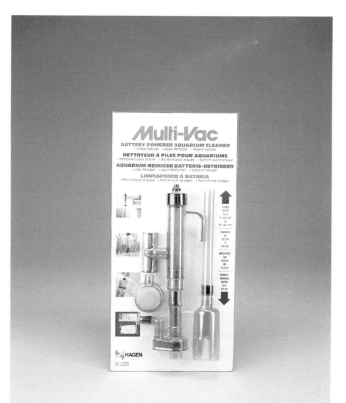

There's no way around it; you are going to *have* to siphon your gravel. The Multi-Vac will help make this chore a little more pleasant. Photo courtesy of Hagen.

charcoal, or filter pads, or hair rollers, you will need to follow the manufacturers recommendations about maintenance of your filter. It is best, however, to: 1. Clean your filter only with lukewarm water (not too thoroughly) to keep your bacterial cultures going. 2. Clean your filter and change your media a few days after you do a water change and gravel wash. The established bacteria in the filter are best able to accommodate the extra load of stirred up wastes and a clean filter is less effective at this time. Besides, if you want to get the last bit of use out of your media, change it after it has trapped the extra waste.

Good water quality is a must for spawning and healthy fry production. Photo by A. Kapralski.

Foods and Feeding

One of the great pleasures of keeping fishes is feeding them, and mbuna will not be disappointing on this count. Healthy mbuna have voracious appetites. They are quite happy to eat just as often as you are willing to feed them. When you see an mbuna that isn't eating, most likely its sick, harassed, or mouthbrooding. And even when holding eggs, some females manage to get a few small bites in past the larvae.

Several small meals a day are more beneficial to your mbuna than one or two large feedings. One or two large feedings encourages them to overeat which could, in turn, contribute to bloat for they will definitely stuff themselves until their stomachs are distended and it *looks* as if they have bloat. In nature they graze almost constantly so they are getting small quantities of food all day long. This is not usually practical for those of us who must spend time away from our aquaria during the day, but several small feedings will net us healthier fishes.

Pseudotropheus aurora. The female essentially fasts during the three weeks she mouthbroods. Good feeding before breeding will help her get through this stressful period. Photo by A. Kapralski.

Facing page: The rasping teeth of mbuna are specialized for grazing on algae. Photo by Dr. Herbert R. Axelrod.

THE VEGETARIAN MENU

Because of their natural diet of auf*wuchs*, you will notice that your mbuna pick at everything in the aquarium. They are following their natural grazing instincts and for this reason are happier in an aquarium that has a good growth of algae.

In most aquaria they will quickly graze back whatever algae that naturally occurs. If you have a spare tank in the house, it is easy to culture some extra algae for their gustatorial pleasure. Set up a small aquarium with some fist-size rocks, add

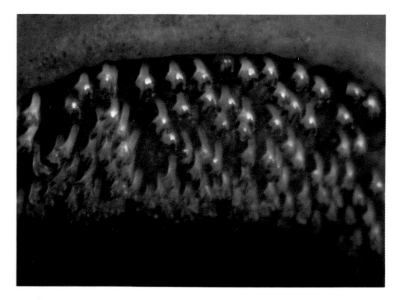

some aquarium water, plenty of light, and presto, instant fish food. When you get a good growth of algae on a rock, trade it for one in your aquarium and you will see your fishes immediately start to scrape the algae off the

Flake foods, like Nutrafin Staple Food, is an important part of a well-balanced diet for fishes. Photo courtesy of Hagen.

Facing page: There is a full line of frozen foods with recipes that include the proper ratio of carbohydrate, protein, fat, vitamins, and trace elements necessary for the health of your fish. Photo by B. Degen.

new rock.

Other good greens for mbuna include boiled spinach, crushed peas, frozen and re-thawed lettuce, and of course vegetable-based commercial foods. Sp*irulina*-based foods are excellent. Frozen food blocks for marine fishes such as angels, tangs, and other marine vegetarians will also send them into a feeding frenzy.

MEATS

While mbuna are not fussy, for the meaty part of their diet live brine shrimp (*Artemia salina*) are tops. Perhaps part of the reason mbuna are so fond of brine shrimp is that they, too, dine on algae, giving the mbuna their meat and vegetable courses in one handy bite-size

morsel. Although live brine shrimp are first choice with any fish, for our convenience brine shrimp are also available in frozen and freeze-dried forms.

Bloodworms are rich; therefore, they should be fed only in small quantities as a treat. Tubificid worms are not recommended for mbuna as they have been implicated as factors in Malawi bloat. The mbuna are not really crazy about them

Brine shrimp.

Blackworms.

Tubifex worms.

Freeze-dried krill.
Photo by I. Français.

anyway.

Larger mbuna will cheerfully devour small goldfish and guppies, but feeding live fishes will make them more aggressive. They get "used" to the idea of being predators and could easily "mistake" smaller tankmates for food. Live fishes are not necessary for their health or well-being, so encouraging them to be more aggressive is probably not a great idea.

We can never emphasize enough the value of feeding live foods to your aquarium fishes. They are often thought of as the

Four treats in one package.
Photo by I. Français.

kinds of insects there are in the world to get some idea of the kinds of live foods available. Many of these insects will send your fishes into a feeding frenzy. Just make very sure that when you do feed the fishes fresh insects that there is no possibility that they could have been contaminated with insecticides.

"haute cuisine" of the aquarium, but in truth they should really be considered the "meat and potatoes." Live foods give your fishes a zest for life that just can't be duplicated.

There is tremendous variety on the live food menu. Just imagine how many different

Flies are an easy-to-catch treat. (And it's always so satisfying to reduce their population.) Get a large jar and put a few drops of the proverbial honey in the bottom, then tape a small funnel (spout down) to the mouth of the jar. The flies can get in with no problem, but they can't find their way out easily. When you have trapped enough flies to suit your purpose, pop the jar, flies and all, into the freezer. The cold will make the flies dopey so you can eventually open the jar without losing all your fish food.

Mbuna will also relish tiny bits of shrimp, clam, mussel, crab, and fish. A small piece of frozen minced cod will last a long time. It's fun to give your fishes a variety of foods and they don't all

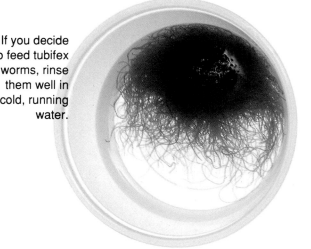

If you decide to feed tubifex worms, rinse them well in cold, running water.

A beautiful pair of *Aulonocara hueseri*. Photo by A. Konings.

Feeding your herd will give you, and them, tremendous satisfaction. Photo by A. Kapralski.

have to come in a package. All it takes is a little imagination and your fishes can have a different dinner every night of the week.

PREPARED FOODS

There is a plethora of fish foods on the market as you will

notice when you visit your pet shop. There are specialized flakes, pellets, granules, freeze-dried live foods, and on and on. Often the dried foods even come with a feeder to measure out the food and keep it in one area of the tank.

FROZEN FOODS

Most pet shops also carry a wide variety of frozen foods like brine shrimp, bloodworms, krill, etc. When you feed frozen foods, break a portion off the frozen block and put it in a fine net, then run some water over the block to 1) thaw the food, and 2) rinse off the water-fouling soup that surrounds the food. Then simply turn the net inside out in the tank and swish the food off the net. Your fishes will love it.

PORTION SIZE

Frequency of feeding is often determined by our schedules rather than our desires. Fishes will survive quite well with one feeding per day. Fishes will thrive and grow well with two, three, or more feedings per day.

Pelleted food is very popular in the hobby because it floats longer than other foods, giving the fish time to feast before it gets lost in the gravel. Photo by I. Français.

The trick is never to overfeed the tank. A good rule of thumb is to feed no more than your fishes will eat in five to fifteen minutes. Observe the fishes and the character of the foods you are offering. If the food floats slowly to the bottom the fishes are going to be able to eat more of it than heavy food that falls immediately to the bottom and into the spaces between the gravel, fouling the water and making a mess. Some foods take a very long time before they sink and you can soon decide just how much is enough for your fishes. Start with a portion that will fit on a dime and see if they eat all of it within the five minutes allotted. If they scoff it up in two minutes, double the ration. If you notice that they are sated and there's food to spare, simply reduce the ration, and remove uneaten foods with a net.

FASTING

Just as it is healthy for most people to fast for a day, it is also good for your fishes. Fasting will give their digestive systems a rest and promote a good appetite the following day.

Facing page: A pair of *Pseudotropheus lombardoi.* The golden fish is the male, the blue is the female. Photo by G. Medla.

Fish Health

We all want our fishes to stay healthy and live for a long time, but a fish like any other living creature is subject to illness, injury, and, sharing the eventual fate of all living things, will someday perish. However, every trauma need not be fatal to the hardy mbuna you are keeping in your aquarium. Even though these little animals sometimes go belly-up for no apparent reason, many of the common ailments suffered by mbuna are easily avoided. Let us discuss some of the problems you could encounter with the health of your fishes and propose

solutions.

While it is laudable to know all the cures for the various illnesses, there is one nostrum that is effective for most maladies—prevention. How does one help prevent fish disease? By being a good aquarist.

STRESS

Just about every fish undergoes considerable stress before it is safely ensconced in your home aquarium. Stress is a major factor of disease and death in your fishes. The most obvious stressful situations are those that cause fright, discomfort, or pain. Handling and capture, while sometimes necessary, should be avoided, since they are primary sources of stress.

Territorial behavior is another source of stress in fishes. It has been determined that the aggressive behavior of dominant fishes can cause severely frayed fins and darkened bodies in the chronically stressed victims. This is especially true of mbuna. If you notice one fish hanging out in a corner constantly and being harassed by another whenever it moves, you know this

Facing page: These two fish are both *Pseudotropheus zebra*. There are many different color morphs of mbuna, not only by sex but also by location within the lake. Photo by A. Konings.

fish is under stress. Either rearrange the rockwork or move the fish to another tank; otherwise, the fish will be dead in a matter of days or hours.

Confinement, anesthesia, bad water quality, transport, and changes in water temperature are all stress-producing factors in fishes. These are things to be avoided whenever possible. While no fish can reach our tanks without undergoing some stress, remember that this is not good for your fishes and will certainly cause problems (just like in humans) if permitted to continue. Studies have shown that proper conditioning is one valuable way to reduce the effects of stress, another valuable method is to ensure that care is taken to minimize the number of stress inducers at one time, such as handling combined with a change in temperature, pH, or poor water quality.

Whether our new fishes are wild or cultivated, they have been captured, stored, crowded, chilled, heated, and transported several times. They are usually underfed (if fed at all) and only achieve

Facing page: *Pseudotropheus zebra*, female and male. The female is the golden fish and the male is the blue and black fish. Photo by A. Konings.

temporary comfort in the dealer's tank. So when we get our prizes home, it is necessary to bear in mind that these poor creatures have been through quite a lot and need to be treated with consideration. When you first bring them home, and before you release them, float the plastic bag in the aquarium to equalize the temperatures and slowly mix the aquarium water with the bag water in order to equalize the chemistry. Your first feeding attempts may be frustrating, not

Facing page: Two *Pseudotropheus zebra*, vastly different in pattern and color intensity. Photo by A. Konings.

necessarily because the fishes are sick, they are just disoriented, and even though hungry they may not be too interested in food.

POISONING

Poisoning is a common cause of death in fishes, even before disease processes have had time to manifest themselves. Chlorine and chloramine must be removed from the water. It is simple enough to age the water to remove the chlorine, and treat with neutralizer (sodium thiosulphate) for chloramine. Another poison that is sometimes overlooked is actually produced by the fishes themselves. The ammonia and nitrite building up daily in your tank can be just as poisonous as DDT and the only treatment for this is prevention. Perform the routine water changes and you will not have to worry about your crystal clear but biologically filthy water weakening your fishes and making them targets for any passing pathogen.

OXYGEN DEFICIENCY

Oxygen deficiency is also contributory to disease processes in our fishes. Regardless of what type of fishes we keep, oxygen is required for respiration. Warmer water has a lower oxygen-carrying ability than cooler water.

The causes of the diseases to which fishes are prone are many, but in general they are caused by bacteria, parasites, viruses, and fungi. It is almost impossible for an aquarist to make an absolutely accurate diagnosis without considerable knowledge, experience, and equipment, but there are some conditions that can be assumed from symptoms.

Facing page: More color morphs of *Pseudotropheus zebra*. No wonder the systematists have such a hard time naming the different species of mbuna! Photo by A. Konings.

DISEASES

Dropsy

Dropsy is a bacterial condition that makes a fish look as if its scales are standing out. Normally healthy fishes are immune to this disease, but they can contract the disease if other fishes in the tank are infected. There is no real cure for this disease, but antibiotics can be tried in an isolation tank.

Ich

Ichthyophthiriasis, commonly called "ich," is a disease caused by a protozoan that burrows into the skin of the fish and causes a little white blister. At first the fish rubs its body on rocks, plants, and other objects in the tank, as if scratching

The distinctive profile of a fish with dropsy. Photo by H.-J. Richter.

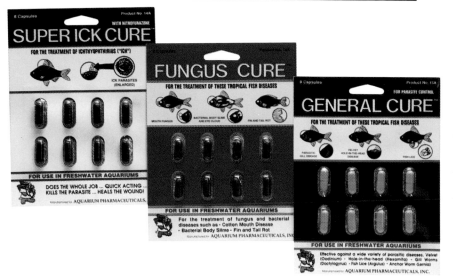

Aquarium Pharmaceuticals has a full line of fish cures. The package literature will help you treat the ailment. Photo courtesy of Aquarium Pharmaceuticals.

itself. Then you can see small white spots on the fins and later on the body of the fish.

Ich is very contagious and if present in one tank can be transferred to another by using the same net, etc. for both. So, as with most diseases, hygiene is a good preventative measure. Wash your hands after working with one tank and before going to another. Use separate nets for each tank. Never trade plants from a diseased tank to a healthy tank. The same goes for fishes. If a fish looks healthy, but has been in a tank with problems, don't put it into a healthy tank. Chances are it is carrying trouble, so cure the whole tank before transferring anything from it to another.

The therapy for ich is to elevate the temperature of the water to about 85 or 86°F. This speeds up the life cycle of the parasite and kills the young. Malachite green, available in many packaged ich cures, is the treatment of choice. Use exactly according to package directions.

Malawi Bloat

Malawi bloat is an internal bacterial infection, not yet suitably explained as to cause, but thought to be a consequence of feeding mbuna too much meat. Being primarily vegetarians, the mbuna are unable to digest meat well. Tubificid worms seem to be especially associated with the condition. The fishes swell up with fluids in their abdominal cavity, and eventually die. They sometimes respond to Furan-based antibiotics, but not always.

Pop-Eye

Pop-eye (Exophthalmia), is easy to identify as the eye becomes enlarged and stands out in the socket. This can affect

Ulcer formation.
Photo by H.-J.
Richter.

Costia
infestation.
Photo by H.-J.
Richter.

one or both eyes. The cause appears to be gas bubbles behind the eye, or possibly a bacterial condition as it sometimes follows mechanical injury. Salt baths are preferred for the treatment of this condition.

Saprolegnia

Saprolegniasis is a fungal infection that attacks injured fishes only. The infected fishes have white fuzzy areas on the body. The treatment for this condition is also malachite green according to package directions. Progressive salt baths are also used with favorable results.

Shimmies

Shimmies is a condition rather than a disease. The fish swims but doesn't get anywhere, wagging its body from side to side. Ich, indigestion, chills, etc., are possible causes of this condition. A water change and elevation of the temperature usually correct this malady.

Tuberculosis

Tuberculosis is a wasting disease and affected fishes look sick. They are hollow-bellied, emaciated, listless, and hold their body crooked. Sometimes the skin will show open lesions. When a fish is this sick, destroy it in a humane manner to save it unnecessary suffering.

Fish can sometimes be afflicted with more than one condition. This unfortunate is suffering from shimmies and a parasitic infestation. Photo by H.-J. Richter.

Fortunately, piscine tuberculosis is not contagious to humans, but it will spread among fishes, so remove victims as soon as possible.

Velvet

Velvet (*Oodinium*) is a parasite that attacks the gills and skin of affected fish. It is recognized by a yellow-brown film that starts near the dorsal fin and spreads to the rest of the body in a velvet-like covering. Acriflavine, methylene blue, or formalin are used in treating this disease.

Pop-eye. Photo by H.-J. Richter.

SALT BATHS

A progressive salt bath is performed in a hospital tank that contains no plants or decorations. Salt is added in the proportion of two level teaspoons per gallon of water on the first day. Dissolve the salt in water before adding it to the fish tank. The second day add two more teaspoons per gallon, and one teaspoon per gallon on the third day. Marine salt is preferred over table salt. When the fish is cured it should be gradually reacclimated to fresh water before being returned to the community tank.

HOSPITAL TANK

The hospital tank is a necessary accessory when you have an outbreak of any kind of illness in your community tank. Adding medications to a well-established tank because one or even several fishes exhibit signs of disease is not a well-advised thing to do. You never want to dose healthy fishes. Some of the medications used in the aquarium require that some or all of the water be totally changed, which could be difficult in the community tank. Also, the disease could become well-established in the tank in the time it takes to attempt a cure.

The large eggs of *Cyathochromis obliquidens* are being laid by the female while the male waits to fertilize them. Photo by A. Kapralski.

Breeding

Part of the proof of your skill in aquaristics comes when your fishes decide to reproduce. Aquarium hobby magazines and books are largely dedicated to reports of breeding fishes, as well as finding them and keeping them, but to breed a new or difficult species is considered the pinnacle of success.

Fortunately, mbuna are among the easiest fishes to breed and unlike so many other freshwater aquarium fishes, will happily do so in the community aquarium. Provided they have the correct conditions and you provide safe havens for the fry, you will soon have several generations coexisting together in the same aquarium.

Unless you know quite a lot about fish genetics, do not attempt to mix and match your fishes. Only breed the best, and two fish of the same species, whether wild-caught or commercially developed. Hybrids are generally worthless. Even if the fry are attractive, the next generation will be hopeless. If you are breeding conspecifics, your chances of having a good spawn are that much better. Try fooling around with Mother Nature and you will probably get garbage. It might seem that crossing those two good-looking fish of different species may produce nice babies, but believe me, the result will never equal the quality of either of the parents.

To ensure that your fry have the very best chances for survival, there are certain things you should do for them. First, choose the very best parents. Condition them well. Make sure they are healthy. Give them the very best water conditions. Give them warmth and light. Give them enough room to grow to their optimal potential. In return for all this care and devotion, your fry will grow up into beautiful fish that you can be proud of.

The fish you choose to breed must be young and healthy. Do not breed immature fish, nor old fish. An immature fish will not produce good fry, nor will it develop properly itself. When you breed immature females, you are reversing the progress that has already been made with this strain and producing small, underdeveloped fish. Your chances of having a spawn worth saving are dramatically increased if the fish are in tip-top condition.

Facing page: The female picks up the eggs in her mouth and will incubate them in her buccal cavity. Photo by A. Kapralski.

WHAT IS A MOUTHBROODER

All mbuna are mouthbrooders. A mouthbrooder is a fish that incubates its eggs in its mouth. There are varying degrees of mouthbrooding. Some fishes brood the eggs and spit out sac fry. Some fishes lay and fan their eggs on the substrate for a time before picking them up, but all mbuna are considered biologically "advanced" by virtue of their mouthbrooding behavior. Mouthbrooding is an advanced breeding behavior in that it affords excellent protection for eggs and fry. It doesn't tie the fish down to one breeding site. The fish can lay the eggs, pick them up, and move on, carrying their fry with them. Without exception, mbuna are maternal mouthbrooders, although some Labeotropheus males have been known to pick up the eggs and hold them for a time, but cases of male mbuna brooding eggs to term are virtually unknown.

SEXUAL DIMORPHISM

Sexing mbuna is something of an acquired art. Many species are strongly

Facing page: A spawning pair of *Pseudotropheus tropheops*. The male is the handsome blue fellow. Photo by A. Kapralski.

sexually dimorphic, as for example *Melanochromis auratus*, in which mature males are silver and black, and females and juveniles are black, white, and gold. In other species the differences are more subtle, as in many morphs of *Pseudotropheus zebra*, for example the cobalt morph, in which both sexes are a bright blue; however, there are still some subtle differences if you have a discerning eye. Males are usually somewhat more intensely colored than females and generally have more numerous and more distinct anal fin ocelli, or egg spots.

In P*seudotropheus lombardoi*, the females are blue, and the males are gold. This is the reverse of most other *Pseudotropheus* sp., where you have blue males and gold females. *Labeotropheus* spp. often have orange or yellow females and the males are blue. In those with OB morphs, the OBs are often females. Male OBs do occur and they are called "marmalade cats." The easiest way to tell them from the females is that there is a metallic blue overlay to the males. Again, it's subtle and it helps to choose from a group of adult fish but it is

Facing page: There can be no doubt who the female is here. She's the one laying the eggs. Photo by A. Kapralski.

impossible to tell with juveniles.

In some mbuna you can tell the sexes right at birth, as with the tangerine zebra, color morphs of P. *zebra*. The females will be gold or orange and the males are a deep midnight blue color. When fully mature, this may not be absolute but most often this is the case.

If you are still uncertain, observe their behavior. If you have a group of fish of the same species and size, and put them in an aquarium, the males are going to be dominant, or at least the dominant fish will probably be a male. Unfortunately, it doesn't really help tell which of the other fish are females and which are subdominant males. Of course, there's always the time when you find a fish with eggs in its mouth, this being the best indicator that it is a female. Also, there are often ambiguous fish, such as the dominant female, who will sometimes fool you behaviorally, but never in the presence of the male.

It is also possible to sex the fish in the hand with a magnifying lens. Net the fish, turn it over, and examine the genital region. In females, there are

Facing page: Two male *Pseudotropheus* (*P. zebra* above, *P.* "Zebra Red Dorsal" below). Photos by Ad Konings.

Two male *Labidochromis freibergi* showing slightly different profiles. Photo by Dr. Herbert R. Axelrod.

usually two distinct pores visible; in males, the anterior pore is small and may appear to be absent. Experienced breeders of African cichlids often call this the "one hole, two hole" method.

CONDITIONING THE PARENTS

The pre-spawning diet is very important. Special pre-spawning feeding is called "conditioning." This is necessary if you wish to have a worthwhile spawn. Doubtless the fishes will go ahead and breed without all this special treatment, but remember, the goal here is a good, healthy clutch of fry, and one that will be worth raising. It would be very disheartening to go to the trouble to save fry and end up with misshapen, puny fish. To properly condition your fishes, feed often and well with live food for about two weeks. You will be able to see the difference. They will be robust and full of vitality. These are the kind of fishes you want to breed. During the conditioning period it is best to keep the sexes separate if possible. You don't want any accidents.

Other factors of the conditioning of your breeders are light and warmth. Gestation time is reduced in bright light. By the same

Facing page: *Pseudotropheus zebra.* Photo by A. Roth.

token, a warmer tank, in the upper range of ideal temperatures, will help bring the embryos to maturity faster.

In the community tank, you will be conditioning every fish in the tank when you are conditioning a pair for spawning. It would be ludicrous to expect that just the two fish you have your eye on are going to get the live brine shrimp you intend for them. In reality, you should be conditioning your fishes all the time. Once mbuna reach adulthood, they are usually getting ready to spawn, have just spawned, or are carrying fry.

Live food is the

This mouthbrooding female albino *Pseudotropheus zebra* is quite at home in her flowerpot cave. Photo by L. Weiner.

The female mouths the male's egg spots to stimulate the release of sperm. Photo by A. Kapralski.

absolute best way to condition your fishes for spawning. Again, live brine shrimp, and light feedings of other live foods, often, will quickly get your fishes into peak spawning condition. Couple this with a rather large water change and, if your fishes are anywhere near ready, they will spawn if the water quality and temperature are favorable.

THE SPAWNING

The best possible case scenario when breeding mbuna is to have one dominant male and a harem of three or more females. Of course, if anything should happen to the lone male, you would have to wait for another male to reach maturity before setting out on your breeding program.

As your fish prepare to spawn, the males will become more colorful than the females. This heightened coloration is called their "nuptial dress." The males will also become more territorial than usual. They will start to dig around their caves, setting up housekeeping for the laying of the eggs. The females as well will change, becoming more rotund, sometimes on only one side of the body.

Your mbuna will happily spawn in the community aquarium, but your fry will have a

The male waits to fertilize the newly laid eggs. Photo by A. Kapralski.

The female picks up the fertilized egg. What looks like another egg on the anal fin of the blue male is actually his "egg spot." Photo by A. Kapralski.

far better chance at survival if you are able to provide a spawning tank for the parents. The spawning tank need only be 15 or 20 gallons with a sponge filter, light, heater, thermometer, substrate, and rocks...a smaller, cozier approximation of the community tank with identical water conditions.

If you can transfer the female(s) first and condition her/them in the spawning tank, so much the better, but we certainly realize that on the first go it is not always possible to (a) catch the female, (b) tell

A rearing tank should have a sponge filter and cover for the fry to feel secure.

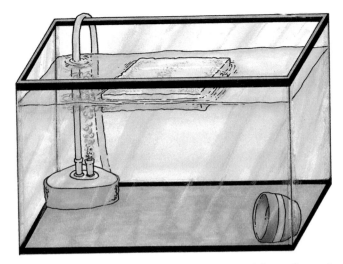

A bare-bottomed breeding tank will make it easy to siphon off uneaten food and waste.

which one is the female, (c) get her out of the community tank before she's already brooding eggs.

When spawning is about to begin, with an excited male and a female in readiness, the male will usually dart out to meet her as she passes through his territory, tense his body, and perform a ritualized dance, spreading his fins and trembling. He will swim all around the female and lure her into his chosen spawning site. After several "dances," during which he waves his anal fin with its egg spots in her direction,

The female will lay about thirty eggs, a few at a time. Photo by A. Kapralski.

Each time the male fertilizes the eggs...

...the female picks them up in her mouth before laying the next batch.
Photo by A. Kapralski.

she will hopefully follow him and begin laying her eggs.

INCUBATION OF THE EGGS

The female picks up the eggs continuously during spawning. The female lays a few eggs, the male fertilizes them, the female picks the eggs up and holds them in her mouth, lays a few more, and so on. Because of their advanced reproductive behavior, mbuna lay very few, large eggs, about 30, as opposed to some other primitive fishes that may scatter literally thousands of eggs only to have two survive to maturity.

If you know that your fish have spawned in a spawning tank, remove the male. He will only be a pest to the brooding female and one more predator when the fry arrive.

More often than not, all this reproductive behavior has taken place when you were not looking and your first indication that the female is carrying eggs is when you notice one fish hanging back at feeding time. What, a mbuna not eating?! Mouthbrooding females usually keep to themselves and don't eat when they have a mouth full of eggs. (Although I have known a few who manage to slip a few tiny morsels past the larvae.) As the days pass, her gular region will protrude more and more, and in

Note the design of the rocks in this scene. The small cave is ideal for this spawning pair of *Pseudotropheus acii*. They don't even notice the fish on the floor below. Photo by A. Kapralski.

The larvae of *Pseudotropheus aurora* stripped prior to release from the buccal cavity of the mother. Photo by A. Kapralski.

light-colored fishes, you will actually be able to see the eyes of the embryos under her chin as dark spots. Sometimes a little tail will even protrude from her lips.

Mbuna brood their eggs for about three weeks. The fry hatch after about 10 to 12 days, but are retained in the female's mouth until they are mature enough to make it on their own. The fry are usually about $1/4$ inch long when they are released. In most species even after the term fry have been released, the female will take the fry back into her mouth when danger threatens...usually for about the first week or so post-release.

Some breeders prefer not to let the female brood to term and they strip the females of their young about 15 days post-spawning. They will catch the female in a net, pry open her mouth with a toothpick, and make her spit out the sac-fry. The fry are spit out into a fine-meshed net that is placed in about an inch of water over a sponge-filter outlet. Here they tumble in the bubbles and develop as well as the fry that were carried to term by the female. The advantage to the breeder in this method is that the valuable fertile female won't have to starve for three weeks. Reproduction takes a

The fry of *Pseudotropheus macrostoma* are taken back into the mother's mouth for protection. Photo by U. Werner.

great deal of parental energy and the sooner the breeder can get the female back on food and into condition, the sooner she will be ready to breed again.

It is not at all unusual to find that all your females are brooding at the same time, that you have two or three spawns of different sizes, and that the dominant male is starting to court his daughters. Now aren't you glad you didn't overstock that tank when you set it up?

CARE OF THE FRY

Once the fry have been released and you notice the tiny fish darting out from their crevices, you will want to provide the right food for the young. Yes, they choose territories right from the start, and they are very comical, appearing for a morsel, then disappearing as if attached to a tight

Newly hatched young from the mother's mouth. Photo by G. Marcuse.

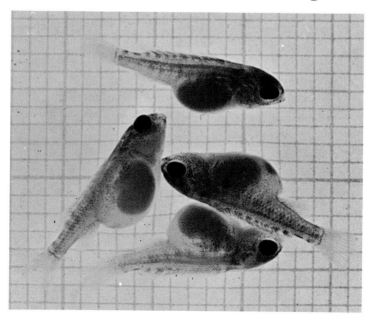

rubber band. Newly hatched brine shrimp are best for young mbuna, but don't be surprised if the first time you see your fry in a community tank is when they come out for their share of flake food.

Albino *Aulonocara* sp. Photo by T. St. John.

The Mbuna Themselves

The popularity of mbuna has led to the publication of many books about them, such as the excellent text shown here.

There are ten commonly recognized genera of mbuna and within these ten genera, literally hundreds of species, color morphs, varieties, and types. The generic descriptions below will give you an idea what to expect from your fishes.

GENUS: CYATHOCHROMIS

Cyathochromis obliquidens is the sole member of the genus. It has a very strong resemblance to the fishes of the genus *Pseudotropheus* with the exception of it's teeth. It is widely distributed in the intermediate zones in shallow water.

GENUS: *CYNOTILAPIA*

Cynotilapia afra and *Cynotilapia axelrodi* are the two most well-known species of *Cynotilapia* of which there are a total of about ten species.

Cynotilapia spp. should be kept by themselves in a large tank because of their fierceness. In the lake, *Cynotilapia* tend to form large schools, at times mixing with schools of *Pseudotropheus zebra*, which they resemble. They are primarily plankton feeders,

Pseudotropheus sp. Photo by A. Kapralski.

Top: *Cynotilapia afra* variety. Photo by Dr. Herbert R. Axelrod.
Bottom: Female *Cynotilapia axelrodi*. Photo by Dr. Herbert R. Axelrod.

Top: *Cynotilapia afra.* Photo by Dr. Herbert R. Axelrod. **Bottom:** *Cynotilapia* sp. Photo by Dr. Herbert R. Axelrod.

Top: *Cynotilapia afra.* Photo by Dr. Herbert R. Axelrod. **Bottom:** *Cynotilapia* sp. Photo by Dr. Herbert R. Axelrod.

Top: *Cynotilapia* sp. Photo by Dr. Herbert R. Axelrod. **Bottom:** Male *Cynotilapia axelrodi*. Photo by Dr. Herbert R. Axelrod.

though they also dine on *aufwuchs*, crustaceans, and small fishes.

GENUS: *GENYOCHROMIS*

Genyochromis mento is commonly known as the Malawian scale-eater, and is very aggressive with a tendency to eat the scales and fins of other fishes. They have the nasty habit of lurking at the bottom of the aquarium and lunging upward for a nip at any passing fishes. They are even attracted to fighting fishes, picking up lost scales and attacking the distracted warriors. For this reason, it is inadvisable

Genyochromis mento. Photo by Arkadi/Vanoff.

to keep them in a community setting. They do not, however, require scales and fins in their diet, and will happily adjust themselves to normal aquarium food, including flake food. Luckily, they are not attracted to the scales and fins of conspecifics. There are many color morphs of this fish, leading some to believe that there may actually be more than one species within the genus.

GENUS: *GEPHYROCHROMIS*

Gephyrochromis lawsi is a loner. Even the males are not very territorial and the females live alone. Uncommon on the rocky shores of the lake, they are most often found in the intermediate zones below 24 feet.

GENUS: *IODOTROPHEUS*

Iodotropheus sprengeri is a non-territorial mbuna that is usually found in pairs or small groups. They are small, reddish-brown fish with a purple-violet cast to each scale. They are known to indulge in some species in-fighting, but on the whole are less aggressive than many of the other mbuna species. This species is known in the hobby under the common name, rusty cichlid.

Top: *Iodotropheus sprengeri*. Photo by K. Paysan. **Bottom:** *Gephyrochromis moori*. Photo by Dr. Herbert R. Axelrod.

Top: *Pseudotropheus* sp. Photo by Dr. Herbert R. Axelrod. **Bottom:** *Gephyrochromis lawsi.* Photo by Dr. Herbert R. Axelrod.

Top: Female *Labeotropheus fuellborni*. Photo by G. Medla. **Bottom:** Female *Labeotropheus fuellborni*. Photo by G. Medla.

The red-top zebra, *Pseudotropheus zebra*. One of the colorful morphs of *P. zebra* and very popular in the hobby. Photo by Photo by Dr. Herbert R. Axelrod.

GENUS: *LABEOTROPHEUS*

Labeotropheus fuelleborni is one of the most peaceful of the mbuna. *Labeotropheus* spp. are most easily identified by their overhanging snout, which gives them a "Roman nose" look.

Labeotropheus trewavasae, or red-top trewavasae, is considered one of the most colorful cichlids in the hobby. A beautiful fish, it is, however, quite aggressive and dominant males will completely overwhelm females and subordinate males unless special measures are taken. Provide a large aquarium and many, many hiding places!

Labeotropheus trewavasae, red-top morph. Photo by H.-J. Richter.

Top: *Labeotropheus fuellborni*, female. Photo by Dr. W. E. Burgess.
Bottom: *Labeotropheus fuellborni*, male. Photo by A. Roth.

Top: *Labeotropheus trewavasae*, female. Photo by A. Roth. **Bottom:** *Labeotropheus trewavasae*, male. Photo by A. Roth.

GENUS: *LABIDOCHROMIS*

Labidochromis is a large genus with many species. *Labidochromis vellicans* was one of the earliest mbuna to be exported for the hobby, but has fallen into relative obscurity with the arrival of so many new and more colorful fishes. *Labidochromis* spp. are remarkable in that their jaws are narrow and their snouts are longer and more pointed than the average mbuna.

GENUS: *MELANOCHROMIS*

The *Melanochromis melanopterus* species-complex differs from most mbuna in that their stripes are horizontal rather than vertical, and most attractively so. Also interesting is the reversal of pigmentation patterns in males and females (male is dark where female is light, and visa versa) of some of the species. They are slender, elongate fishes with wide mouths and slightly thickened lips. These fishes are usually only mildly territorial in the lake and the many species are very colorful and excellent in the aquarium.

The *Melanochromis* 'heterogenous' species-group lack the distinct longitudinal banding found in the *M. melanopterus* spp. group. Most of the 'heterogeneous' group

Top: *Labidochromis* sp. Photo by Dr. W. E. Burgess. **Bottom:** *Labidochromis vellicans*. Photo by Dr. Herbert R. Axelrod.

Top: *Melanochromis vermivorus.* Photo by A. Roth. **Bottom:** *Labidochromis* sp. Photo by Dr. W. E. Burgess.

are deeper-bodied fishes.

While most *Melanochromis* spp. are very weakly territorial in Lake Malawi, it should be remembered that the crowded conditions in the aquarium will not bring out their best behavior. Any mbuna should be considered aggressive and territorial in an aquarium and offered safe sanctuaries from restless tankmates.

GENUS: *PETROTILAPIA*

Members of the genus *Petrotilapia* are the largest of the mbuna, measuring about 5 inches at adulthood. Their mouths are always slightly open and the many tiny teeth are easily seen, almost giving the appearance that their inner lips are lined with velcro. Males are most interested in defending their territories from other males of the same species, and will vigorously attack intruders.

GENUS: *PSEUDOTROPHEUS*

The genus *Pseudotropheus* contains many, many species and it sometimes seems more are being added every day. The following are some of the more available and identifiable species.

P. elongatus, as the name suggests, is an elongated fish, easily

Both of these fish are *Petrotilapia* spp. Photo by A. Konings.

Top: *Pseudotropheus tropheops.* Photo by Dr. W. E. Burgess. **Bottom:** *Pseudotropheus novembfasciatus.* Photo by K. Paysan.

identified by its long, slender body. Both sexes are blue with broad vertical dark bands, but the males are much brighter blue with a dark head.

P. lanisticola lives in empty snail shells. As the fish grows it must look for larger and larger shells. They are quite nasty and should be provided with many shells and other holes to help maintain the peace. These fish are generally either bluish or yellowish, with the bluish fish being the breeding male.

P. livingstoni lives mainly in sandy areas and uses the shells of the snail *Lanistes nyassanus* as its refuge and spawning area. When keeping this species, some large, empty apple snail or escargot shells will make it feel right at home.

P. lombardoi females and juveniles are a bluish color with black barring, with the adult males taking on a striking golden color with black barring. The males are very aggressive within their own species, but generally leave other species alone.

P. tropheops has many varieties based on beautiful color forms and location within the lake. *P. tropheops* was one of the first mbuna known to science. They are recognized by the profile of the snout which descends steeply.

Pseudotropheus zebra varieties are common in the hobby, very easy to keep, and an excellent beginner's fish, although very quarrelsome if not given its own space. There are many color morphs and varieties, particularly the very attractive OB morphs. *P. zebra* males in the wild defend territories centered around a cave where they spawn or takes refuge. Females, youngsters, and sub-dominant males are social, travelling together in schools.

Aulonocara is given honorary status as an mbuna for its life-style The origins of the genus *Aulonocara* are decidedly different from those of the "other"

Pseudotropheus lombardoi. Photo by K. Paysan.

Top: *Pseudotropheus tropheops.* Photo by G. Axelrod. **Bottom:** *Pseudotropheus tropheops.* Photo by G. Axelrod.

mbuna, eliminating *Aulonocara* from the group from a strictly taxonomic point of view. Bearing in mind that mbuna is simply a Chitonga word meaning rock-dwelling fish, and not a scientific classification, and that *Aulonocara* are indeed rock-dwelling fish, their inclusion in the mbuna club is not out of order.

Aulonocara ethelwynnae. Photo by A. Konings.

Above: *Aulonocara* sp. hansbaenschi, male. Photo by A. Konings.
Below: *Aulonocara* sp. jumbo blue, female. Photo by A. Konings.

References in boldface refer to illustrations.